WORKBOOKS IN CHEMISTRY

SERIES EDITOR

STEPHEN K. SCOTT

WORKBOOKS IN CHEMISTRY

Beginning mathematics for chemistry
Stephen K. Scott

Beginning calculations in physical chemistry
Barry R. Johnson and Stephen K. Scott

Beginning organic chemistry 1
Graham L. Patrick

Beginning organic chemistry 2
Graham L. Patrick

WORKBOOKS IN CHEMISTRY

Beginning Organic Chemistry 2

GRAHAM L. PATRICK

Department of Chemistry
Paisley University

OXFORD   NEW YORK   TOKYO
OXFORD UNIVERSITY PRESS
1997

Oxford University Press, Great Clarendon Street, Oxford OX2 6DP

Oxford   New York
Athens   Auckland   Bangkok   Bogota   Bombay   Buenos Aires
Calcutta   Cape Town   Dar es Salaam   Delhi   Florence   Hong Kong
Istanbul   Karachi   Kuala Lumpur   Madras   Madrid   Melbourne
Mexico City   Nairobi   Paris   Singapore   Taipei   Tokyo   Toronto
and associated companies in
Berlin   Ibadan

Oxford is a trade mark of Oxford University Press

Published in the United States
by Oxford University Press Inc., New York

© Graham L. Patrick, 1997

A catalogue record for this book is available from the British Library

Library of Congress Cataloging in Publication Data
(Data applied for)

ISBN 0 19 855936 4

Typeset by the author
Printed in Great Britain by
Progressive Printing UK Ltd
Leigh-on-Sea, Essex

Workbooks in Chemistry: Series Preface

The new *Workbooks in Chemistry Series* is designed to provide support to students in their learning in areas that cannot be covered in great detail in formal courses. The format allows individual, self-paced study. Students can also work in groups guided by tutors. Teaching staff can monitor progress as the students complete the exercises in the text. The Workbooks aim to support the more traditional teaching methods such as lectures. The format of the Workbooks has been evolved through experience and discussions with students over several years. Students benefit through the Examples and Exercises that provide practice and build confidence. University staff faced with increasing class sizes may find the Workbooks helpful in encouraging 'self-learning' and meeting the individual needs of their students more efficiently. The topics covered in the early Workbooks in the Series will concentrate on background support appropriate to the early years of a Chemistry degree, including mathematics, performing calculations, and basic concepts in organic chemistry. These should also be of interest to students who are taking chemistry courses as part of other degree schemes, such as biochemistry and environmental sciences. Later Workbooks will be designed to support material typically encountered in later years of a Chemistry course.

Contents

Introduction

The following self-learning text has been prepared to help you in your studies of organic chemistry. By working through the various sections, you will be led through some of the basics of organic reactions and reaction mechanisms. You will also get plenty of practice at problem solving and answering the sorts of questions you might get asked in an examination. In fact, this is probably one of the greatest advantages of using this text in that it gives you model answers for all the problems and questions asked and lets you see what examiners expect of you when they set a question. Too often, students fail to do themselves justice in an exam because they have either not put down enough in their answer or have strayed from the point.

The text is split into various sections as follows:

Section 1   Nucleophiles and Electrophiles
Section 2   Reactions and Mechanisms
Section 3   Classification of Reactions
Section 4   Acid/Base Reactions
Section 5   Electrophilic Addition
Section 6   Electrophilic Substitutions 1
Section 7   Electrophilic Substitutions 2
Section 8   Nucleophilic Additions
Section 9   Nucleophilic Substitutions of Carboxylic Acid Derivatives
Section 10   Nucleophilic Substitutions of Alkyl Halides
Section 11   Reactions Involving Acidic Protons Attached to Carbon—Aldehydes and Ketones
Section 12   Elimination Reactions of Alkyl Halides and Alchohols
Section 13   Reductions and Oxidations
Section 14   Radical Reactions

In each section, you will get a short bit of theory, then a question. An empty box will be provided for your answer and the correct answer follows on. To get the best out of this text, you should read your lecture notes and textbook on the subject you wish to study, then find the relevant section in the self-learning text. Try to answer the question before looking at the answer and use a pencil so that you can reuse the self-learning text at a later date. If you find it difficult to resist the temptation of 'sneaking a look' at the answer, you could always paste pieces of paper over the answers, which you can then fold back as and when you wish to look at the answer. Don't worry if you don't get the correct answers the first time you tackle a particular section. You would be exceptional if you did. The important thing is to try the questions and to come up with your own answers. Once you have identified your mistakes and you repeat the section at a later stage, you will make fewer errors. Your goal is to keep repeating the section until you don't make any errors at all and can answer the questions in the way shown. In this way, you should get all your mistakes and errors ironed out before the exam comes along.

SECTION 1

Nucleophiles and Electrophiles

Nucleophiles and Electrophiles

1.1 Definition of nucleophiles and electrophiles

Most organic reactions involve the reaction between a molecule which is rich in electrons and a molecule which is deficient in electrons. Electron-rich molecules which take part in reactions are called **nucleophiles**. Electron-deficient molecules which take part in reactions are called **electrophiles**. Nucleophiles will have a specific region of the molecule which is electron-rich. This region is called the **nucleophilic centre**. Electrophiles will have a specific region of the molecule which is electron-deficient. This region is called the **electrophilic centre**.

So far so good, but how do you recognise nucleophiles and electrophiles? In this section, we shall answer that question.

1.2 The nucleophilic and electrophilic properties of ions

Let us first consider the following two ions. Do you think they are electron-rich or electron-deficient? If so, which is which? Which is the nucleophile and which is the electrophile? Can you identify any electrophilic centres or nucleophilic centres in the molecules?

Answer

This should be easy. If a molecule is negatively charged, it must be electron-rich. It is therefore a nucleophile. Therefore, the hydroxide ion is a nucleophile. The nucleophilic centre will be the atom which has the negative charge and this is the oxygen atom.

The carbocation is positively charged and electron-deficient. It will be an electrophile. The electrophilic centre will be the atom with the positive charge, so this is the carbon atom.

What general rule do you think we could make about the nucleophilic/electrophilic properties of ions and the position of nucleophilic/electrophilic centres in ions? Complete the following.

> Negatively charged ions will be electron .... and will behave like .... philes. The
>
> .... philic centre of the ion will be the atom bearing the .... charge
>
> Positively charged ions will be electron .... and will behave as .... philes. The
>
> .... philic centre will be the atom bearing the .... charge.

Answer

Negatively charged ions will be electron-**rich** and will behave like **nucle**ophiles. The **nucle**ophilic centre of the ion will be the atom bearing the **negative** charge.

Positively charged ions will be electron-**deficient** and will behave as **electro**philes. The **electro**philic centre will be the atom bearing the **positive** charge.

Here are some simple examples of nucleophiles and electrophiles. Identify which molecules are nucleophiles and which are electrophiles. Identify the nucleophilic/electrophilic centre in each case.

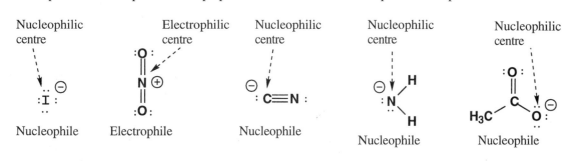

Answer

All these molecules have a negative or positive charge and so it is easy to identify whether they are nucleophiles or electrophiles and to pinpoint the location of the nucleophilic/electrophilic centre.

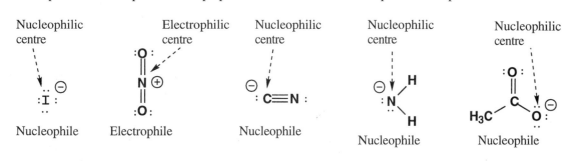

1.3 Nucleophilic and electrophilic properties of neutral inorganic compounds

Recognising nucleophilic and electrophilic centres in neutral molecules is not as simple as recognising them in ions, but there are several guidelines which we can use to help us. Let us first look at the water molecule and see if we can identify any nucleophilic or electrophilic centres.

There are no positive or negative charges in this molecule, but are there any atoms with a partial charge? Such partial charges are possible if there are any polar bonds.

Answer

You should be aware that O–H bonds are polar covalent. This is because the oxygen atom is significantly more electronegative than the hydrogen atom. As a result, the oxygen has a greater share of the electrons in the O–H bond and is slightly negative. Since the hydrogen has a smaller share of these electrons it must have a slightly positive charge.

Taking into account the polar bonds in water, can you identify electrophilic and nucleophilic centres?

Answer

Since water has both nucleophilic and electrophilic centres, you may well ask whether the molecule will react as an electrophile or a nucleophile. Well, the answer is that it can react as either, depending on what it reacts with. For example, if water is treated with an electrophile, it will behave like a nucleophile. On the other hand, if water is treated with a nucleophile, it will behave like an electrophile.

As an illustration of this, consider the following reactions. Identify whether water is reacting with a nucleophile or an electrophile in each of these reactions, and then determine whether water is reacting as an electrophile or as a nucleophile.

(a)

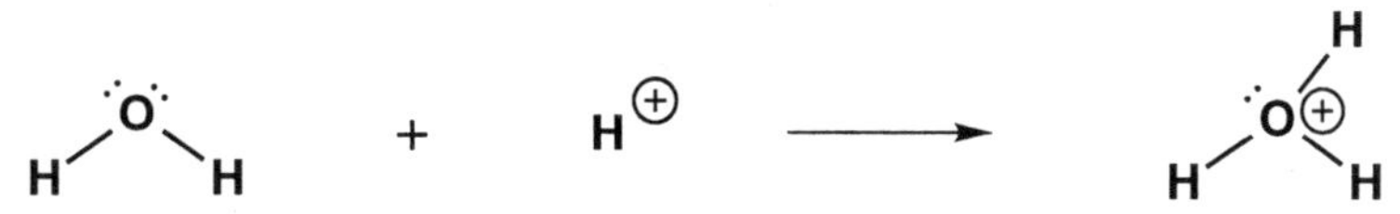

(b)

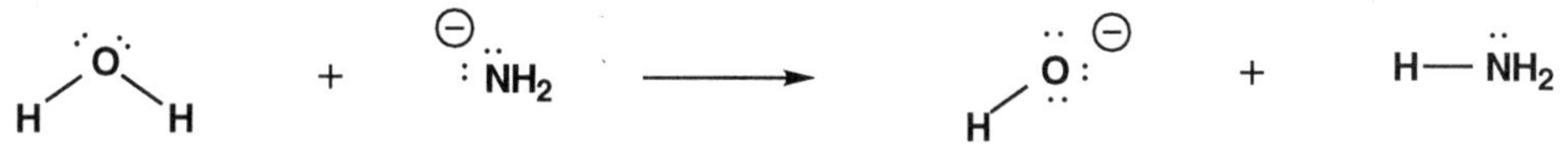

Answer

(a)

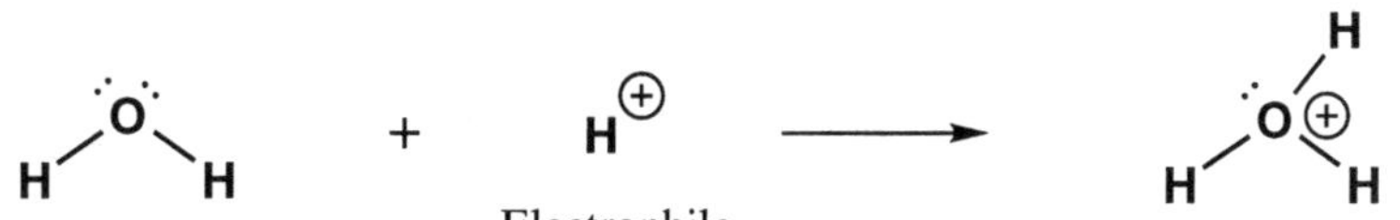

Electrophile

In reaction (a) the proton is positively charged and is an electrophile. It will therefore react with a **nucleophilic** centre on water. The nucleophilic centre on water is the oxygen atom, which uses one of its lone pairs to form a bond to the proton. Therefore, water reacts here as a nucleophile.

(b)

Nucleophile

In reaction (b), the NH_2^- ion is a nucleophile and will therefore react with an **electrophilic** centre on water (i.e. one of the hydrogen atoms). The nucleophilic centre on the NH_2^- ion is the nitrogen atom. This atom uses one of its lone pairs of electrons to form a bond to a hydrogen on water. The product is ammonia and a hydroxide ion. Water reacts here as an electrophile.

The following molecules all contain polar covalent bonds. Identify these and hence identify possible nucleophilic and electrophilic centres in each molecule.

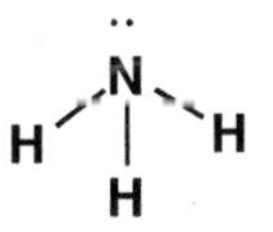

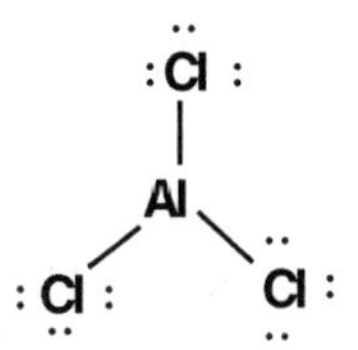

Answer

The polar covalent bonds lead to partial charges as shown below. Those atoms with a slightly negative charge are electron-rich and are therefore nucleophilic centres. Those atoms with a slightly positive charge are electron-deficient and are therefore electrophilic centres.

Although ammonia has electrophilic and nucleophilic centres, it usually reacts as a nucleophile. By contrast, hydrogen fluoride and aluminium chloride prefer to react as electrophiles and very rarely react as nucleophiles. Water appears to be in between since it can react as a nucleophile or as an electrophile. What conclusions can you draw regarding the relative nucleophilic strengths of neutral nitrogen, oxygen and a halogen?

Answer

The evidence suggests that halogen atoms are only weakly nucleophilic, whereas nitrogen atoms are strongly nucleophilic. Oxygen appears to be somewhere in between. (Note that here we are discussing **neutral** halogen, oxygen and nitrogen **atoms** incorporated in a molecule. We have already seen that negatively charged **ions** such as $^{\ominus}OH$, $^{\ominus}NH_2$ or $Br^{\ominus}$ are good nucleophiles.)

What conclusions can you draw regarding the relative electrophilic strengths of the hydrogens in ammonia, water and hydrogen fluoride?

Answer

Since hydrogen fluoride prefers to react as an electrophile, the hydrogen in HF must be a strong electrophilic centre. In contrast, the hydrogens in ammonia are very weakly electrophilic. The electrophilic strength of the hydrogens in water is somewhere in between.

Can you see any trend which relates the nucleophilic strengths of the atoms N, O and F to their position in the periodic table?

Answer

As one moves to the right across the periodic table from nitrogen to oxygen and then to fluorine, the atoms become less nucleophilic.

Is the relative strength of the nucleophilic centres explained by the relative electronegativities of the atoms involved?

Answer

It would appear not. Electronegativity increases from nitrogen to fluorine, so one would expect fluorine to be more electron-rich than oxygen or nitrogen. Yet it appears to be the weakest nucleophilic centre.

In fact, the relative nucleophilic strengths of these atoms can be explained if we look at the products which would be formed if these atoms *were* to act as nucleophiles. If they are to act as nucleophiles with say a proton, these atoms must provide the electrons for the new bond. As a result, they will gain a positive charge as shown below. Which atom is able to tolerate a positive charge the best? Which tolerates it the least?

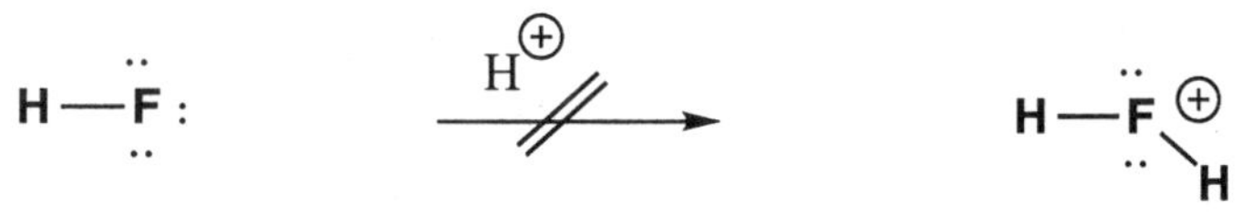

cont.

Answer

If hydrogen fluoride is to act as a nucleophile with an electrophile like hydrogen, then the fluorine atom will gain a positive charge. The fluorine atom is electronegative and does not tolerate a positive charge. Therefore, this reaction does not take place. Oxygen is less electronegative and is able to tolerate the positive charge slightly better, such that an equilibrium is possible between the charged and uncharged species. Nitrogen is the least electronegative and tolerates the positive charge even better, shifting the equilibrium to the salt.

The same argument can be used in reverse when looking at the electrophilicities of the hydrogens in HF, H_2O and NH_3. In this case, reaction with a strong nucleophile or base would generate the following anions.

Which are the most stable anions and why? How does this affect the electrophilicities of the hydrogens in hydrogen fluoride, water and ammonia?

Answer

The most electronegative atom will tolerate the negative charge the best. This is the fluorine atom. Oxygen is able to tolerate a negative charge, but nitrogen is not so tolerant of a negative charge. If the anion is stable, it will be easily formed and hence the hydrogen which is lost will be strongly electrophilic. This is the case for HF. If the anion is unstable (e.g. the $^{\ominus}NH_2$ anion), it is not easily formed and the hydrogen (in ammonia) is a very weak electrophile. Nitrogen anions are only formed with very strong bases.

The relative stability of anions is also influenced by the size of the anions. Thus, within the halogen family, stability follows the order $I^{\ominus} > Br^{\ominus} > Cl^{\ominus} > F^{\ominus}$. This means that the electrophilicity or acidity of the corresponding acids follows the order HI > HBr > HCl > HF.

In conclusion, we can identify nucleophilic and electrophilic centres in a molecule by identifying polar covalent bonds. However, some nucleophilic and electrophilic centres are usually too weak to react as nucleophiles or electrophiles. Hence, halogen atoms in a molecule rarely react as nucleophiles, while hydrogen atoms attached to nitrogen rarely react as electrophiles.

1.4 Nucleophilic and electrophilic properties of alkanes

We shall now look at various organic molecules to see whether they have nucleophilic or electrophilic centres and whether they are likely to react as electrophiles or as nucleophiles. Consider the following alkane. Has it any electrophilic or nucleophilic centres?

Answer

There are no charges in the molecule and there are no polar covalent bonds. There are no electrophilic or nucleophilic centres, and the molecule is unreactive to most chemical reagents.

This is true for all alkanes. C–H and C–C bonds are covalent bonds. Consequently, there are no electrophilic or nucleophilic centres in these molecules. Since most reagents are nucleophiles or electrophiles, they will not react with alkanes since there are no electrophilic or nucleophilic centres to react with.

1.5 Nucleophilic and electrophilic properties of amines

Amines contain a nitrogen atom which we have already seen is a good nucleophilic centre. Amines can have N–H and N–C bonds. We have already looked at the N–H bond. Do you think N–C bonds are polar, and if so to what extent?

Answer

Nitrogen is immediately to the right of carbon in the periodic table, so we would expect it to be more electronegative. However, since they are so close to each other in the table, the difference is slight and we can usually ignore the N–C bonds when looking for electrophilic and nucleophilic centres.

As far as amines are concerned, the nitrogen will always be a nucleophilic centre since it has a lone pair of electrons which makes it electron-rich. Any N–H bond present in amines will be polar and the hydrogens will be very weak electrophilic centres. Identify the nucleophilic or electrophilic centres in the following amines and determine whether these molecules will behave as nucleophiles or electrophiles.

Answer

The nitrogen atoms have a lone pair of electrons and are electron-rich. Therefore, they are nucleophilic centres. The hydrogens directly attached to nitrogen are slightly positive due to the polar nature of the N–H bond. They are therefore very weak electrophilic centres. The dominant feature of amines will be the nucleophilic nitrogen and so amines will usually behave as nucleophiles.

1.6 Nucleophilic and electrophilic centres in alcohols and ethers

We have already seen that oxygen with its two lone pairs of electrons can act as a nucleophilic centre. We have also seen that hydrogens attached to oxygen are electrophilic centres. C–H and C–C bonds are not polar and so these do not provide any nucleophilic or electrophilic centres. What other type of bond is present in alcohols and ethers such as the following?

Answer
Carbon–oxygen bonds are present in both alcohols and ethers.

What can you say about the polarity of this bond? Is it likely to result in an electrophilic centre?

Answer

Oxygen is further to the right of carbon than nitrogen so the C–O bond should be more polar than the C–N bond. Since oxygen is more electronegative, the carbon atom should be slightly electron deficient and will be an electrophilic centre. This turns out to be the case. However, the carbon atom is not a strong electrophilic centre and alcohols and ethers are more likely to be nucleophilic through the oxygen atom.

Identify the electrophilic and nucleophilic centres in the following alcohols and ethers.

$H_3C\!\!-\!\!\ddot{O}\!:$
$\quad \backslash CH_3$

$H\!\!-\!\!\ddot{O}\!:$
$\quad \backslash CH_3$

$H\!\!-\!\!\ddot{O}\!:$
$\quad \backslash CH_2\!-\!CH_2$
$\qquad\qquad \backslash CH_3$

Answer

$H_3C\!\!-\!\!\overset{\delta-}{\ddot{O}}\!:$
$\overset{\delta+}{} \backslash \underset{\delta+}{CH_3}$

$H\!\!-\!\!\overset{\delta-}{\ddot{O}}\!:$
$\underset{\delta+}{} \backslash \underset{\delta+}{CH_3}$

$H\!\!-\!\!\overset{\delta-}{\ddot{O}}\!:$
$\underset{\delta+}{} \backslash \underset{\delta+}{CH_2}\!-\!CH_2$
$\qquad\qquad \backslash CH_3$

The oxygens have the partial negative charge and are electron-rich. Therefore, they are the nucleophilic centres. The hydrogens and carbons directly attached to oxygen will be electron-deficient and are therefore the electrophilic centres.

In alcohols, there are two possible electrophilic centres—the carbon atom directly attached to oxygen and the hydrogen directly attached to oxygen. Which do you think will be more electrophilic?

Answer

The O–H bond will be more polar than the C–O bond, so the hydrogen atom will be more electrophilic. However, the hydrogen atom is still a very weak electrophile in alcohols and so alcohols generally react as nucleophiles through their oxygen atom.

1.7 Nucleophilic and electrophilic centres in alkyl halides

Identify the nucleophilic and electrophilic centres in the following alkyl halides.

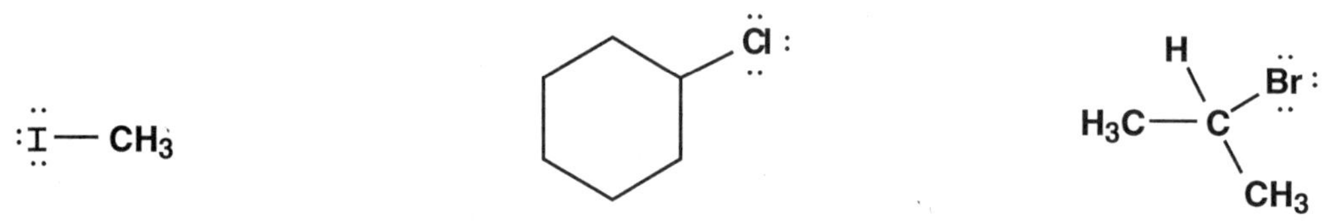

Answer

The halogens are the nucleophilic centres in these molecules. The carbon–halogen bonds are polar bonds due to the electronegativity of the halogen atom. Therefore, the carbon atoms involved in these bonds will be electron-deficient and will be electrophilic.

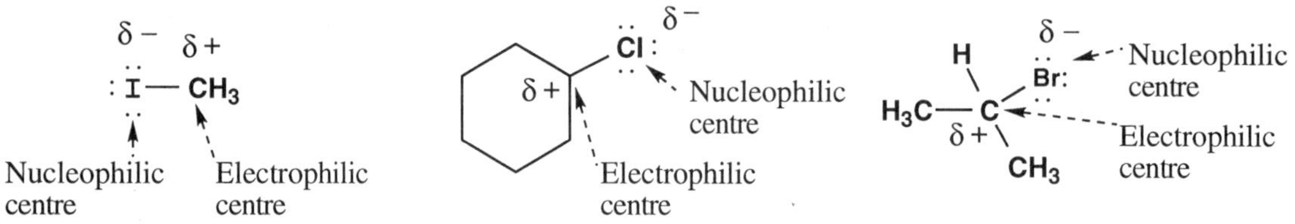

Based on what you have learnt in previous sections, do you think these molecules are more likely to react as electrophiles or as nucleophiles?

Answer

We have already seen that halogen atoms within molecules are poor nucleophiles. Therefore, alkyl halides are more likely to react as electrophiles at the electrophilic carbon centre.

1.8 Nucleophilic and electrophilic centres in aldehydes and ketones

Aldehydes and ketones contain a carbonyl (C=O) group where one bond is a strong sigma (σ) bond and the other bond is a weak pi (π) bond. The following structures are two examples of aldehydes and ketones. Identify any nucleophilic or electrophilic centres which you think might be present.

Answer

We can ignore the C–H and C–C bonds since these are non-polar. There is an oxygen present with two lone pairs of electrons, so this will be electron-rich and can be a nucleophilic centre. The carbon next to the oxygen is the carbonyl carbon. This is electrophilic since the oxygen is electronegative and has a greater share of the electrons in the C=O bonds.

Aldehydes and ketones have a carbonyl bond which has an electrophilic carbon and a nucleophilic oxygen. These molecules can react as nucleophiles or as electrophiles-depending on the nature of the other reactant.

1.9 Nucleophilic and electrophilic centres in carboxylic acids and acid derivatives

We have looked at the nucleophilic and electrophilic centres of aldehydes and ketones. Both these functional groups contain a carbonyl (C=O) group. However, there are other functional groups which include a C=O group. What are they?

Answer

The other functional groups which contain C=O are the carboxylic acids and their derivatives (i.e. acid chlorides, acid anhydrides, esters, amides).

Like ketones and aldehydes, the carboxylic acids and their derivatives have both electrophilic and nucleophilic centres. Identify these in the following examples.

Answer

The oxygens and nitrogens in the above structures are electron-rich and are therefore nucleophilic centres. The carbons of the C=O bonds are electron-poor and are electrophilic. This is because the oxygen at the other end of the C=O bond is more electronegative than carbon and pulls electrons away from the carbon atom.

As you can see, there are quite a few nucleophilic and electrophilic centres in these molecules and consequently several possible sites where a nucleophile or an electrophile could react. We shall look at some of these reactions in a later section.

1.10 Nucleophilic and electrophilic centres in alkenes, alkynes and aromatic compounds

Alkenes, alkynes and aromatic compounds are nucleophilic. In other words, they are electron-rich and prefer to react with electrophiles (molecules which are electron-deficient). Suggest why these molecules are electron-rich and identify the nucleophilic centres.

Answer

All these functional groups have multiple bonds (i.e. they have double or triple bonds between two carbons). Consequently, the area in space between the multiple bonded carbons must be rich in electrons. The nucleophilic centre in these molecules is not an atom, but the bond(s)!

Identify the nucleophilic centres in the following molecules.

Answer

Note that the whole of the aromatic ring has to be considered as the nucleophilic centre. That is because the π electrons are delocalised round the ring.

1.11 Nucleophilic and electrophilic centres in multi-functional molecules

Most interesting molecules in organic chemistry and biochemistry contain more than one functional group. They are called multi-functional as a result. Since they may have several functional groups, they will have several nucleophilic and electrophilic centres. This makes the number of possible reactions quite large and it is part of an organic chemist's skill to be able to identify which of several nucleophilic or electrophilic sites is the most likely site for a particular reaction.

Consider the following molecule and answer the questions below.

(1) Outline and identify the functional groups in the molecule.

cont.

(2) Identify the sites in the molecule where a nucleophile might react.

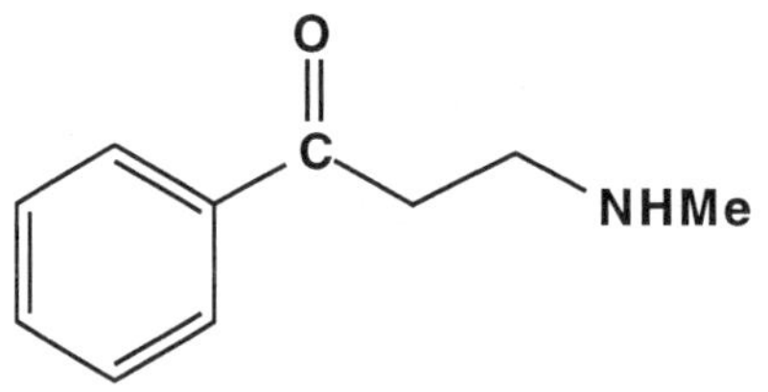

(3) Identify the sites in the molecule where an electrophile might react.

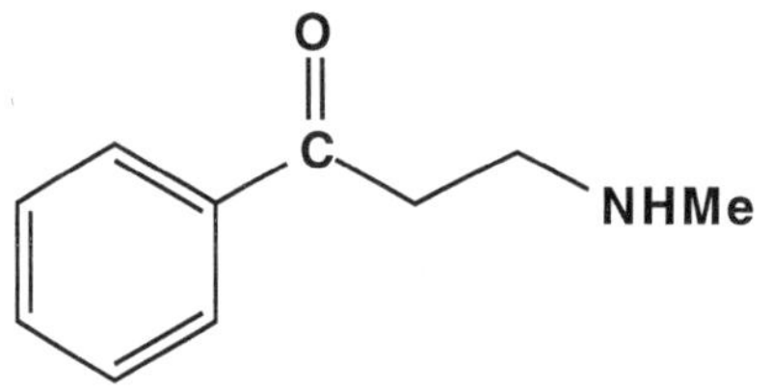

Answer

(1) The functional groups are as follows.

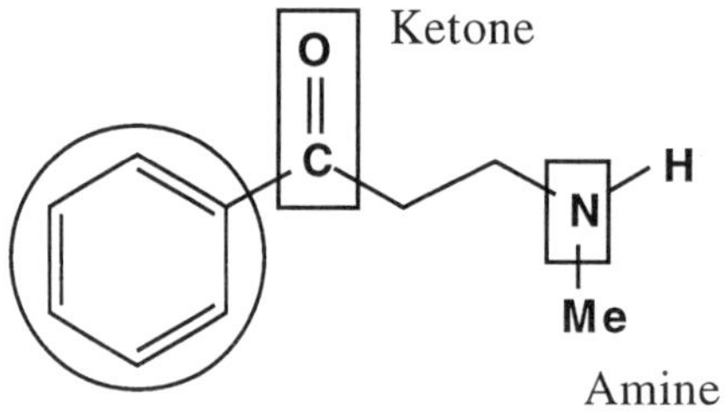

(2) A nucleophile being electron-rich would react at an electron-poor site. Therefore, it would react at electrophilic centres in the molecule. The possible sites for reaction are shown below.

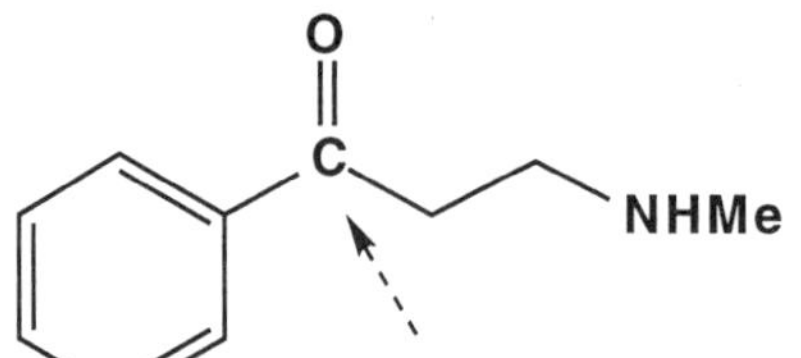

There is only one electrophilic site and therefore only one possible place for a nucleophile to react. (The carbon beside nitrogen is only very weakly electrophilic and can be ignored.)

cont.

(3) An electrophile is electron-poor and will therefore react with nucleophilic or electron-rich centres on the molecule. These are shown below.

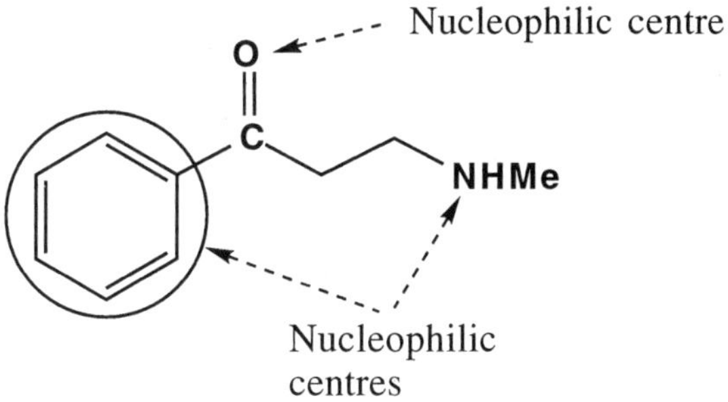

There are three possible places in this molecule where an electrophile can react.

Repeat the above question with the following molecule.

Functional groups

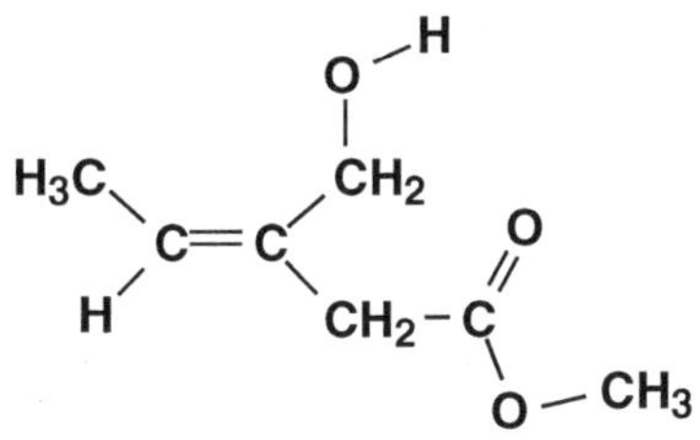

Electrophilic centres

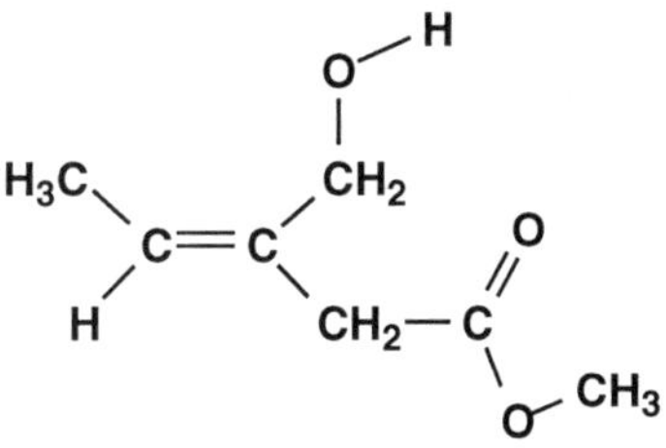

Nucleophilic centres

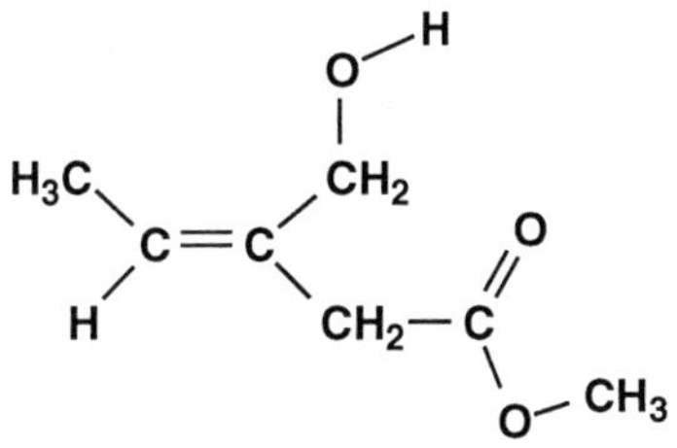

Answer

(1) The functional groups are as shown.

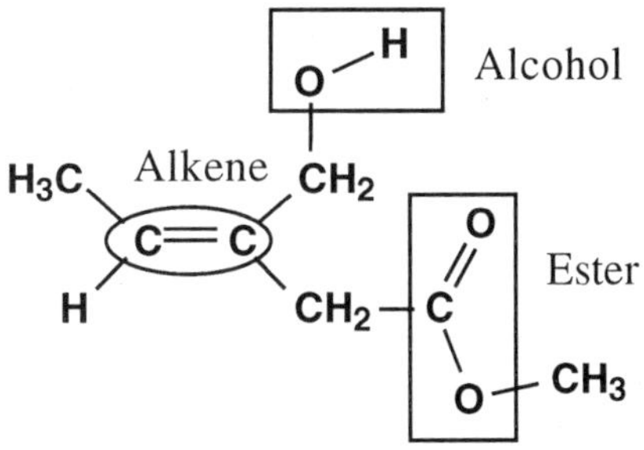

(2) Nucleophiles react at electrophilic centres. There are four such centres in this molecule.

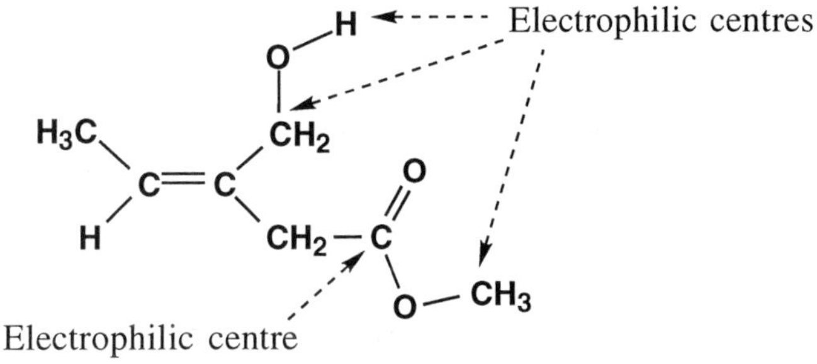

The carbonyl carbon and the OH hydrogen are the most likely centres to be attacked by a nucleophile.

cont.

(3) Electrophiles react at nucleophilic centres. There are several such centres in the molecule.

The following molecule is called procaine and is an important local anaesthetic. Outline and identify the functional groups.

Answer

Aromatic amine Aromatic ring or arene Aromatic ester Amine

Now identify the nucleophilic and electrophilic centres.

Nucleophilic and electrophilic centres

Answer

The nucleophilic and electrophilic centres are the following.

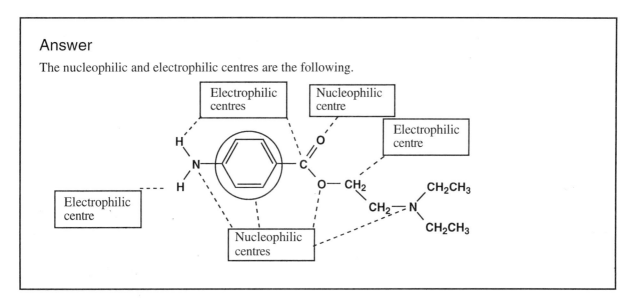

Summary

- Nucleophilic centres in a molecule are electron-rich and can react with electrophiles.

- Electrophilic centres in a molecule are electron-deficient and can react with nucleophiles.

- Negatively charged ions are nucleophiles. The nucleophilic centre is the atom with the negative charge.

- Positively charged ions are electrophiles. The electrophilic centre is the atom with the positive charge.

- Atoms with lone pairs of electrons can act as nucleophilic centres (e.g. N, O, halogens).

- Double bonds, triple bonds and aromatic ring systems can act as nucleophilic centres.

- Hydrogen atoms attached to electronegative atoms are electrophilic.

- Carbon atoms attached to oxygen or a halogen are electrophilic.

- A nitrogen atom is more nucleophilic than an oxygen atom. Halogens within a molecule are only very weakly nucleophilic.

- Hydrogen atoms attached to halogens are more electrophilic than hydrogens attached to oxygen, which in turn are more electrophilic than hydrogens attached to nitrogen.

SECTION 2

Reactions and Mechanisms

Reactions and Mechanisms

2.1 Reactions

You should now be able to identify nucleophilic and electrophilic centres in an organic molecule. However, if you have more than one of these sites, how do you know which site is going to react with a particular nucleophile or electrophile? It is important that chemists can predict whether a reaction will occur and also where it will occur. Otherwise, they would not be able to plan synthetic routes to important compounds. Therefore, organic chemists must 'know their reactions'.

The more reactions that a chemist knows, the easier the job becomes. It's a bit like chess. A grand master has to know the pieces and the moves that can be made before planning a game strategy. As far as an organic chemist is concerned, he/she has to know the molecules and the sort of reactions which can be carried out before planning a synthetic 'game strategy'.

Inevitably, there is a lot of memory work involved in learning these reactions, but there is a logic behind the reactions as well. The most important thing to remember is that most reactions involve electron-rich molecules forming bonds to electron-poor molecules (i.e. nucleophiles forming bonds to electrophiles). The bond will be formed specifically between the nucleophilic centre of the nucleophile and the electrophilic centre of the electrophile.

Consider the following molecules. One is an amine and the other is a ketone. If a chemist reacted these molecules together, do you think a reaction would occur? If it did, what are the reaction centres in each molecule (i.e. where are the electrophilic and nucleophilic centres)?

Answer

The electrophilic and nucleophilic centres in each molecule are shown below.

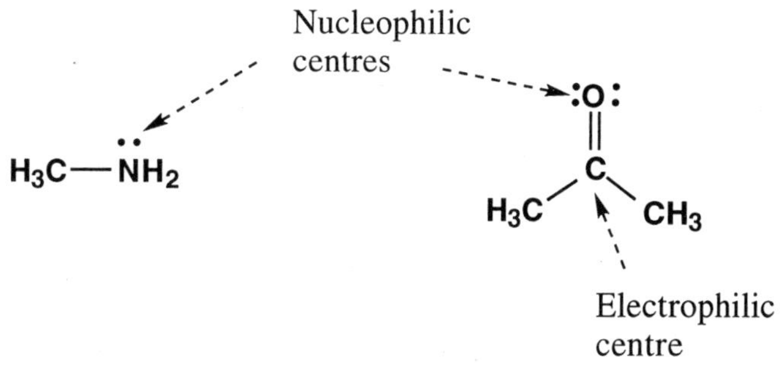

The amine has one nucleophilic centre (the nitrogen) and will therefore 'seek out' an electrophilic centre on the ketone (the carbon shown). We would expect a new bond to be formed between the amine nitrogen and the ketone carbon. This reaction does in fact take place. We will find out what product is formed in a later section.

Repeat the above question with the following amine and alkyl halide.

Answer

The electrophilic and nucleophilic centres are as follows.

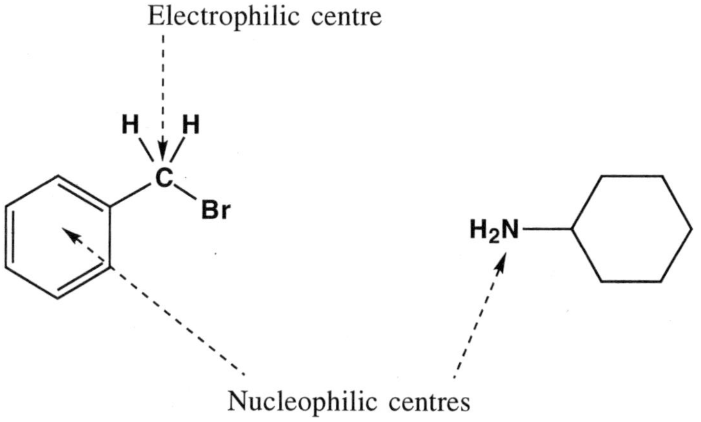

Both molecules have a nucleophilic centre, but only the alkyl halide has a significant electrophilic centre.

cont.

Therefore, the amine will act as the nucleophile and the alkyl halide will act as the electrophile. The bond will be formed between the nucleophilic centre (nitrogen) and the electrophilic centre (carbon). (Note that we have ignored the bromine as a potential nucleophilic centre since it is a very weak nucleophile.)

The actual products obtained from this reaction are shown below. The new bond between the two reaction centres is highlighted.

You will also notice that a hydrogen atom and a bromine atom have left the molecule. We will see why this happens later on.

2.2 Mechanisms

You are now at the stage where you can predict *where* reactions might occur but not what *sort* of reaction will occur. Before we look at that, we need to look at some basic theory on **mechanisms**. A mechanism is the 'story' of how a reaction takes place. It tells us how the starting materials or reagents react together to give the final product.

The mechanism is essential to understanding the reaction. It's a bit like a Sherlock Holmes detective story. At the beginning of the story, we have a crime and we know the suspects. If we flick to the end of the book we would find out the identity of the dastardly villain, but we would have no idea how Holmes solved the mystery (i.e. the mechanism he used). This is hardly satisfactory.

With a chemical reaction, we know what we start off with and we know what we finish with, but to understand the reaction we want to know the story in between. This is the mechanism. The mechanism tells us how bonds are formed and how bonds are broken and in what order things happen.

In a mechanism, we study what is happening to the valence electrons in the molecule since it is the movement and rearrangement of these electrons which leads to a reaction taking place. Let us take as a simple example the reaction between the hydroxide ion and a proton to form water.

The hydroxide ion is the nucleophile and the proton is the electrophile. A reaction takes place between the nucleophilic centre (oxygen) and the electrophilic centre (hydrogen) and water is formed. The mechanism looks at what happens to the electrons and answers questions such as:

- Where did the electrons come from to form the new O–H bond?

- What happened to the third lone pair of electrons on the hydroxide ion? (The oxygen in water only has two lone pairs of electrons.)

These two questions are related and the obvious conclusion is that the third lone pair of electrons has been used to form the new bond to the proton. This means that both electrons for the new bond came from the oxygen. This makes sense since a proton has no electrons. Once the lone pair of electrons is shared between the oxygen and hydrogen, it means that the oxygen effectively 'loses' the equivalent of an electron and the proton effectively gains the equivalent of an electron. As a result, the oxygen loses its negative charge and the proton loses its positive charge. We can show this in a diagram by using a curly arrow.

The arrow is double headed  $\curvearrowright$ (rather than single headed $\curvearrowright$) as it symbolises that both electrons in the lone pair are moving to form a bond to the proton.

Notice that the arrow starts from the source of the two electrons (the lone pair) and points to where the *centre* of the new bond will be formed, not to the proton itself. In some textbooks, you may see the arrow written directly to the proton. Indeed, many chemists have fallen into the habit of drawing it like that, but formally this is incorrect. Arrows should only be drawn directly to an atom if the electrons are both going to end up there as a lone pair.

Remember the following:

- Mechanism arrows must start from the source of electrons (i.e. a lone pair or the middle of a bond).

- Mechanism arrows only point to an *atom* if the electrons are going to become a lone pair on that atom.

- If the electrons are being used to make a new bond, the mechanism arrow must point to where the centre of the new bond will be.

Amines also react with protons. For example, propylamine reacts as shown to give a cation. Draw the arrow to represent the mechanism for this reaction and explain why the nitrogen gains a positive charge.

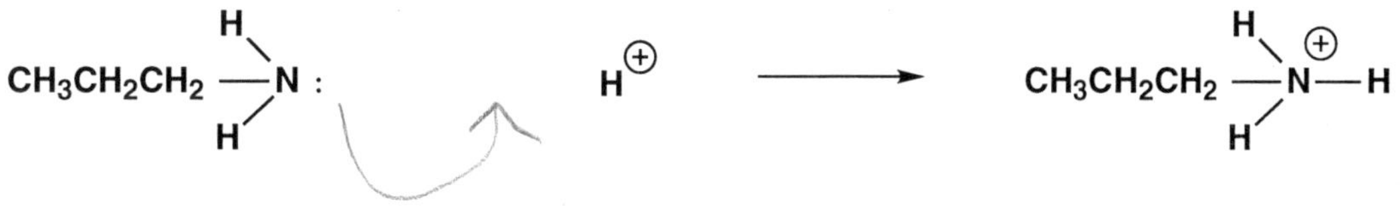

Answer

The reaction takes place at the expected centres. The nitrogen on the amine is nucleophilic and reacts with the proton which is electrophilic. If we study the starting material and the product, we can see that there is a nitrogen lone pair on the former but not on the latter. There is an extra bond in the product and a positive charge on the nitrogen. The obvious conclusion is that the nitrogen's lone pair of electrons has been used to form the new bond, as shown below.

cont.

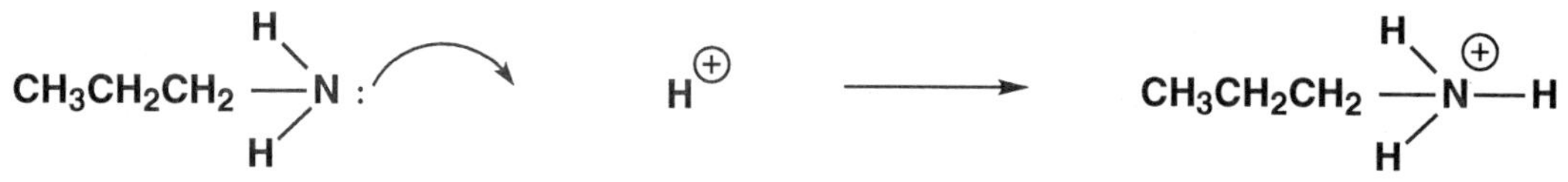

The electrons of the lone pair are now in the new bond and are shared between H and N. Therefore, the nitrogen loses the equivalent of one electron and becomes positively charged.

Reaction of a carboxylic acid with hydroxide ion gives a carboxylate anion, as shown below. Draw a mechanism and explain the appearance/disappearance of charges.

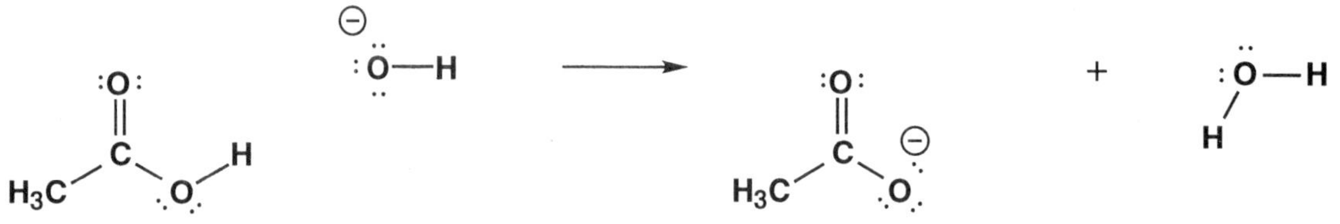

Answer

One of the lone pairs on the hydroxide ion is used to form a bond to the acidic proton. However, hydrogen can only have one bond. Therefore, the bond already present must be broken at the same time. The electrons end up on the carboxylate anion as a third lone pair. The mechanism is as follows.

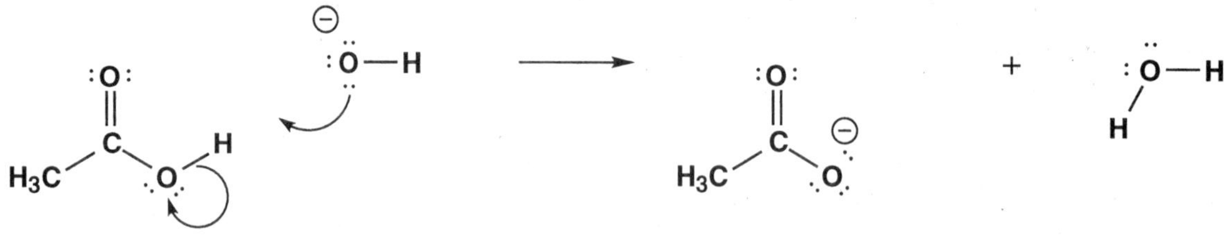

The negatively charged oxygen of the hydroxide ion ends up as a neutral oxygen in water. This is because one of the oxygen's lone pairs is used to form the new bond. Both electrons are shared and the oxygen effectively loses one electron and its negative charge. The oxygen in the carboxylate ion becomes negatively charged since it gains both electrons in the original O–H bond as a third lone pair. It effectively gains an electron.

Note again how the arrows are drawn above. Each arrow *must* start from a pair of electrons. If it is a lone pair of electrons, the arrow starts from there. If the electrons are in a bond then the arrow must start from the middle of the bond. Each arrow must point to where the electrons end up. If the electrons form a new lone pair, then the arrow must point to the atom which bears the lone pair. If the electrons are forming a new bond then the arrow points to where the centre of the new bond will be.

The following diagrams all have something wrong with them. Identify the errors in each case.

(a)

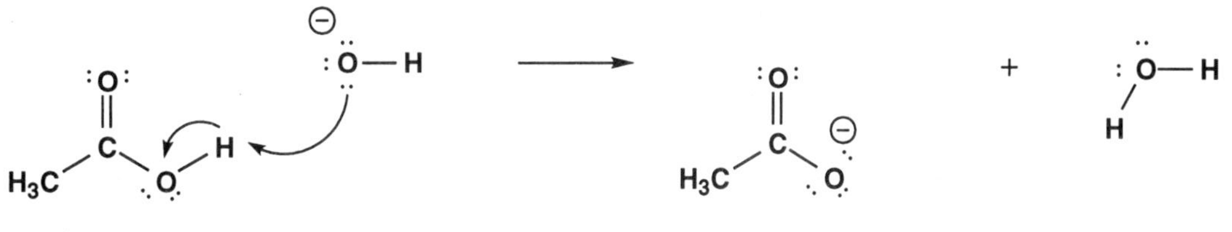

(b)

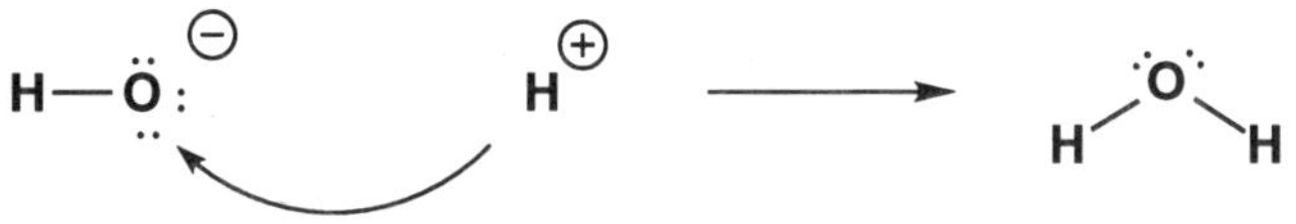

(c)

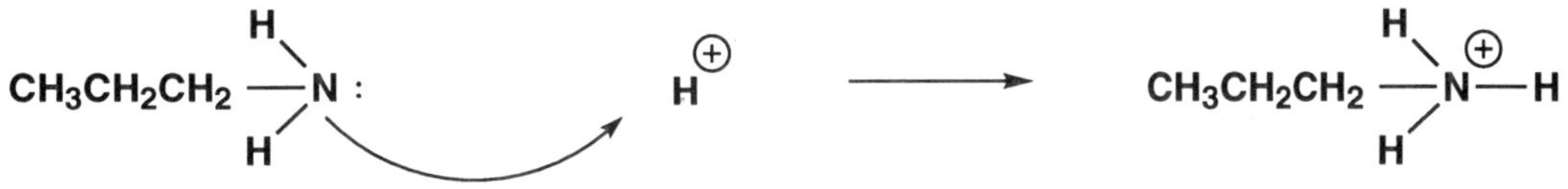

Answer

(a)

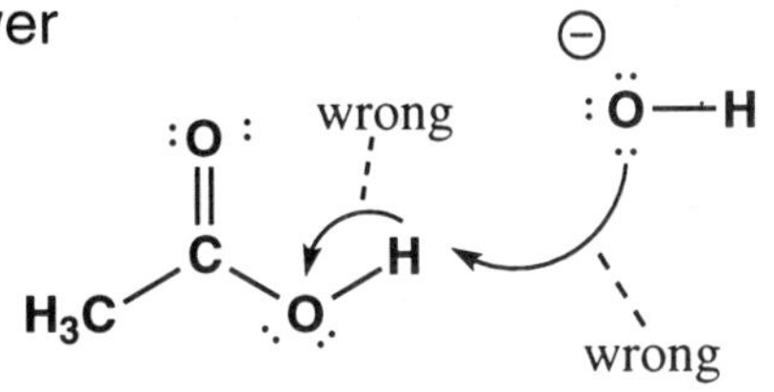

Both arrows are badly drawn.
   The left-hand arrow should start from the centre of the OH bond since it is the electrons in the bond which are moving onto the oxygen.
   The right-hand arrow should not be pointing directly at H but should be pointing halfway between the oxygen and hydrogen (where the centre of the new bond will be formed).

(b)

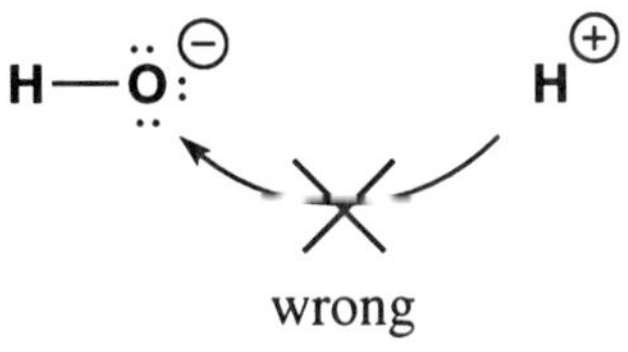

This is one of the commonest mistakes made by students. The arrow is drawn the wrong way. The arrow should be used to show what happens to electrons. There are no electrons on a proton and so the arrow as drawn does not make sense.
   The arrow should point the opposite way to where the centre of the new bond will be formed.

cont.

The correct mechanisms have already been drawn earlier in this section. Draw the correct mechanisms here and check that you got them right by looking at the previous answers (pages 27–28).

Summary

- Most reactions take place between electrophiles and nucleophiles. The reaction involves the formation of a new bond between the nucleophilic centre of the nucleophile and the electrophilic centre of the electrophile.

- The nucleophilic centre provides both electrons for the new bond.

- If the electrophilic centre has its maximum number of bonds, one bond must break as the new bond is formed.

- The mechanism shows which electrons are involved in the making and breaking of bonds.

- Double-headed curly arrows represent the movement of two electrons.

- Arrows used in mechanisms must start from a source of electrons (i.e. a lone pair or the centre of a bond).

- Arrows used in mechanisms will only point to an atom if the electrons end up as a lone pair on that atom.

- If a new bond is being created, the mechanism arrow points to where the centre of the new bond will be formed.

SECTION 3

Classification of Reactions

Classification of Reactions

There are a large number of reactions in organic chemistry but we can simplify the story by grouping these reactions into various categories. To begin with, we can classify reactions as follows:

- acid/base reactions,

- functional group transformations,

- carbon–carbon bond formations.

The first category of reactions are relatively simple and involve the reaction of an acid with a base to give a salt. These reactions are covered in Section 4.

The second category of reactions are methods by which one functional group can be converted into another (e.g. ketone to an alcohol). These reactions can be divided into the following categories:

- electrophilic addition,

- electrophilic substitution,

- nucleophilic substitution,

- nucleophilic addition,

- elimination,

- reduction,

- oxidation.

Specific functional groups are quite 'choosy' about the sort of reactions they will undergo. The following list shows which types of reactions are favoured by specific functional groups.

Reaction category	**Functional group**
electrophilic addition	alkenes and alkynes
electrophilic substitution	aromatic rings
nucleophilic substitution	alkyl halides, carboxylic acid derivatives (acid chlorides, anhydrides, esters, and amides)
nucleophilic addition	aldehydes and ketones
eliminations	alkyl halides and alcohols
reductions	alkenes, alkynes, aromatics, aldehydes, ketones, nitriles, carboxylic acids and their derivatives.
oxidations	alkenes, alcohols, aldehydes.

These reaction categories will be covered in Sections 5–11.

The third category of reactions is extremely important to organic chemistry since they are the reactions which allow the chemist to build up big molecules from smaller ones. These reactions are, in general, the most difficult and temperamental to carry out. In rock-climbing terminology, carbon–carbon bond formations represent the 'crux' to the overall synthesis of a compound. Carbon–carbon bond formations will be covered mostly in later studies of chemistry, but we will see some examples of this type of reaction in this text (e.g. Grignard and aldol reactions).

SECTION 4

Acid/Base Reactions

Acid/Base Reactions

4.1 Acids and bases as electrophiles and nucleophiles

We have already seen that nucleophiles (electron-rich molecules) react with electrophiles (electron–deficient molecules) such that a bond is formed between the nucleophilic centre of the nucleophile and the electrophilic centre of the electrophile.

In the reactions of acids with bases, the acid acts as the electrophile and the base acts as the nucleophile. Before we look at these reactions we have to define what is meant by an acid and a base. There are two definitions—the Brønsted–Lowry definition and the Lewis definition. We shall concentrate on the former and discuss the latter briefly at the end of the section.

4.2 Brønsted–Lowry theory of acids and bases

This is the theory with which you will be most familiar. The Brønsted–Lowry definition of an acid is a molecule which can provide a proton. The Brønsted–Lowry definition of a base is a molecule which can accept that proton.

A simple acid/base reaction is the reaction of ammonia with water.

$$H_2O \quad + \quad NH_3 \quad \rightleftharpoons \quad {}^{\ominus}OH \quad + \quad {}^{\oplus}NH_4$$

Which species is the base and which is the acid?

Answer

Water is losing the proton and is the acid. Ammonia is accepting the proton and is the base.

The reaction is in equlibrium, and if we consider the reverse reaction, the $^{\ominus}OH$ acts as a base and $^{\oplus}NH_4$ acts as an acid.

The following reactions involve an acid and a base. Identify which is which.

$$NH_3 \quad + \quad CH_3CO_2H \quad \longrightarrow \quad \overset{\oplus}{N}H_4 \quad + \quad CH_3CO_2^{\ominus}$$

[handwritten]		*[handwritten] A*

$$\overset{\oplus}{N}H_4 \quad + \quad H_2O \quad \rightleftharpoons \quad NH_3 \quad + \quad H_3O^{\oplus}$$

A	B

$$H_2O \quad + \quad CH_3CO_2H \quad \rightleftharpoons \quad H_3O^{\oplus} \quad + \quad CH_3CO_2^{\ominus}$$

B	A

$$CH_3NHCOCH_3 \quad + \quad H^{\ominus} \quad \longrightarrow \quad CH_3\overset{\ominus}{N}COCH_3 \quad + \quad H_2$$

A	B

$$PhOH \quad + \quad HO^{\ominus} \quad \longrightarrow \quad PhO^{\ominus} \quad + \quad H_2O$$

A	B

$$CH_3NH_2 \quad + \quad H_3O^{\oplus} \quad \longrightarrow \quad CH_3\overset{\oplus}{N}H_3 \quad + \quad H_2O$$

B	A

Answer

$$NH_3 \quad + \quad CH_3CO_2H \quad \longrightarrow \quad \overset{\oplus}{N}H_4 \quad + \quad CH_3CO_2^{\ominus}$$

BASE	ACID

$$\overset{\oplus}{N}H_4 \quad + \quad H_2O \quad \rightleftharpoons \quad NH_3 \quad + \quad H_3O^{\oplus}$$

ACID	BASE

$$H_2O \quad + \quad CH_3CO_2H \quad \rightleftharpoons \quad H_3O^{\oplus} \quad + \quad CH_3CO_2^{\ominus}$$

BASE	ACID

$$CH_3NHCOCH_3 \quad + \quad H^{\ominus} \quad \longrightarrow \quad CH_3\overset{\ominus}{N}COCH_3 \quad + \quad H_2$$

ACID	BASE

cont.

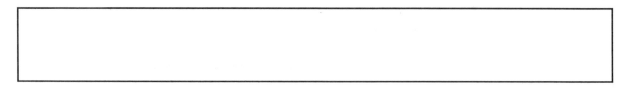

PhOH   +   HO⁻   ⟶   PhO⁻   + H_2O

ACID   BASE

CH₃NH₂   +   H₃O⁺   ⟶   CH₃NH₃⁺   + H_2O

BASE   ACID

4.3 The mechanism of acid and base reactions

Consider again the acid/base reaction between ammonia and water. Ammonia is acting as the nucleophile and water is acting as the electrophile.

$H_2\ddot{O}$:   + $\dot{N}H_3$ ⇌   :$\ddot{O}$H⁻   +   NH₄⁺

The nucleophilic centre in ammonia will form a bond to the electrophilic centre in water. Which electrons are used for that bond?

Answer

The nitrogen uses its lone pair of electrons to form the new bond.

If nitrogen forms a bond to one of the hydrogens on water and nothing else happened, how many bonds would hydrogen have? Is this allowed?

Answer

Hydrogen would have two bonds, which is not allowed.

What must happen at the same time as nitrogen makes its bond to hydrogen?

Answer

The bond between hydrogen and oxygen must break and the electrons making up the bond must move onto oxygen as a lone pair of electrons.

Draw the mechanism to show that.

Answer

Mechanism

The nitrogen atom on ammonia uses its lone pair of electrons to form a new bond to a hydrogen atom on water. As this bond forms, the existing O–H bond in water must break since hydrogen is only allowed one bond.

4.4 Recognising a Brønsted–Lowry acid

A molecule which contains an acidic hydrogen is a Brønsted–Lowry acid. In order to be acidic, the hydrogen must be slightly positive or electrophilic. We looked at what makes hydrogen atoms electrophilic in Section 1. Reread that section if you have forgotten it and state how you would recognise an electrophilic hydrogen in a molecule.

Answer

If hydrogen is attached to an electronegative atom, it will be electrophilic and acidic.

Identify the acidic protons in the following molecules.

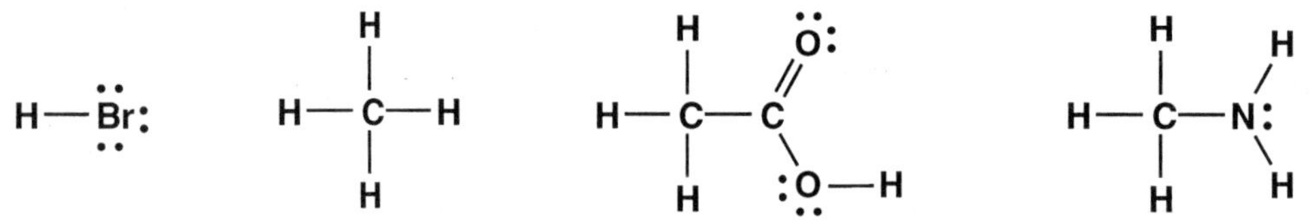

Answer

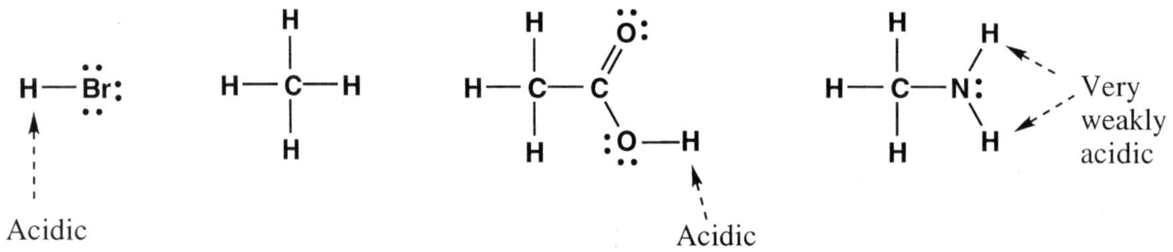

Acidic

Acidic

Very
weakly
acidic

Hydrogens attached to an electronegative atom such as a halogen, oxygen or nitrogen will be acidic. Hydrogens attached to carbon are not normally acidic. However, in Section 11, we shall look at special cases where hydrogens attached to carbon **are** acidic.

If hydrogens attached to oxygen or nitrogen can be acidic, what other functional groups could have acidic protons apart from amines and carboxylic acids?

Answer

If you have looked through all the functional groups which you have met, you should have identified two other functional groups which contain an O–H bond. These are phenols and alcohols.

cont.

Ar—Ö: $\delta-$
Phenol H $\delta+$

R—Ö: $\delta-$
Alcohol H $\delta+$

The other functional groups which contain N–H bonds are primary and secondary amides.

1° amide

2° amide

Although all these molecules have acidic protons, they are not all equally acidic. Compare the following molecules. Which would you expect to have the strongest acidic proton and which the weakest? Explain your choice. (Revise Section 1 if necessary.)

H—F:

Hydrofluoric acid

Very

Ethanoic acid

Quite

Methylamine

Slight.

Answer

Hydrofluoric acid will have the most acidic proton since it is attached to electronegative fluorine. The fluorine strongly polarises the bond to hydrogen such that the hydrogen becomes highly electron deficient and is easily lost. The fluoride ion can stabilise the resulting negative charge.

The acidic protons on methylamine on the other hand are attached to the less electronegative nitrogen. The bond is not polarised so much and the protons are less electrophilic. If the protons are lost, the nitrogen is left with a negative charge which it cannot stabilise as efficiently as a halide ion.

Ethanoic acid will be in between, since oxygen is less electronegative than a halogen but more electronegative than a nitrogen atom.

In the examples above, we have considered hydrogen attached to elements in the same row of the periodic table.

When we come to consider elements of the same family (e.g. the halogens) we find that size plays a role in acid strengths as well. For example, the acid strengths of HF, HCl, HBr and HI increase with the size of the halogen. Nevertheless, all these acids are mostly ionised in water and the electrophilic/acidic proton is more significant than the weakly nucleophilic halide.

By contrast, if you dissolve ethanoic acid (acetic acid) in water, an equilibrium is set up and not all the acidic protons are lost from the carboxylic acid.

If you dissolve methylamine in water, none of the acidic protons are lost at all. The nucleophilic nitrogen is more important than the electrophilic proton and so the molecule behaves as a base instead.

Methylamine can still act as an acid but it has to be treated with a much stronger base than water. Butyllithium could be used, for example (Li^+ is present as a counterion).

4.5 Acid strength and pK_a

Acids can be described as being weak or strong and the pK_a is a measure of this. What does this mean? Let us look at ethanoic acid as an example. If you dissolve ethanoic acid in water, then the following equilibrium is set up.

The reaction is an equilibrium reaction. What is the relationship between the equilibrium constant (K_{eq}) and the concentration of reactants and products?

$$K_{eq} =$$

Answer

$$K_{eq} = \frac{[products]}{[reactants]} = \frac{[CH_3CO_2^-][H_3O^+]}{[CH_3CO_2H][H_2O]}$$

K_{eq} is normally measured in a dilute aqueous solution of the acid and so the concentration of water is nearly constant. Therefore, we can rewrite the equilibrium equation into a simpler form, known as the acidity constant K_a, which includes the concentration of pure water (55.5 M).

$$K_a = K_{eq} [H_2O] = \frac{[CH_3CO_2^-][H_3O^+]}{[CH_3CO_2H]}$$

The acidity constant is a measure of the equilibrium between products and reactants and is also a measure of how acidic a particular acid is. Bearing that in mind would you expect a strong acid to have a low value of K_a or a high value of K_a?

Answer

A strong acid would have a high value of K_a since a strong acid loses protons easily and the equilibrium is mostly over to the products.

The K_a values (in brackets) of a selection of acids are as follows. Place the acids into their order of acidity.

CH_3CO_2H (1.75×10^{-5}) Cl_3CCO_2H (23200×10^{-5}) $ClCH_2CO_2H$ (136×10^{-5})

Cl_2CHCO_2H (5530×10^{-5}) $CH_3CH_2CH_2CO_2H$ (1.52×10^{-5}) $PhCO_2H$ (6.3×10^{-5})

CH_3CH_2OH (1.0×10^{-16}) HCN (6.31×10^{-10}) $PhOH$ (1×10^{-10})

CF_3CH_2OH (3.98×10^{-13}) *p*-nitrophenol (6.31×10^{-8}) $CH_3CH_2NH_2$ $(ca\ 10^{-40})$

CH_3CONH_2 (1×10^{-15}) NH_3 (1×10^{-33})

Answer

The stronger the acid, the better it is at providing a proton and the more the equilibrium is shifted to the right. The more the equilibrium is shifted to the right, the higher the value of K_a. Therefore, the order of acidity is as follows, starting from the strongest acids.

Cl_3CCO_2H (23200×10^{-5}) $> Cl_2CHCO_2H$ (5530×10^{-5}) $> ClCH_2CO_2H$ (136×10^{-5}) $> PhCO_2H$ (6.3×10^{-5}) $> CH_3CO_2H$ (1.75×10^{-5}) $> CH_3CH_2CH_2CO_2H$ (1.52×10^{-5}) $> p$ nitrophenol (6.31×10^{-8}) $> HCN$ (6.31×10^{-10}) $> PhOH$ (1×10^{-10}) $> CF_3CH_2OH$ (3.98×10^{-13}) $> CH_3CONH_2$ (1×10^{-15}) $> CH_3CH_2OH$ (1.0×10^{-16}) $> NH_3$ (1×10^{-33}) $> CH_3CH_2NH_2$ $(ca\ 10^{-40})$

Working with K_a values is often inconvenient due to the awkward numbers involved and it is more usual to measure acidic strengths as pK_a values. The pK_a is the negative logarithm of K_a and gives more manageable numbers.

$$pK_a = - Log_{10} K_a$$

Work out the pK_a values for all the following acids. (The K_a values are above.)

CH_3CO_2H	Cl_3CCO_2H	$ClCH_2CO_2H$
Cl_2CHCO_2H	$CH_3CH_2CH_2CO_2H$	$PhCO_2H$
CH_3CH_2OH	HCN	PhOH
CF_3CH_2OH	*p*-nitrophenol	$CH_3CH_2NH_2$
CH_3CONH_2	NH_3	

Answer

The pK_a value for each molecule is in brackets.

CH_3CO_2H (4.76)	Cl_3CCO_2H (0.63)	$ClCH_2CO_2H$ (2.87)
Cl_2CHCO_2H (1.26)	$CH_3CH_2CH_2CO_2H$ (4.82)	$PhCO_2H$ (4.2)
CH_3CH_2OH (16)	HCN (9.2)	PhOH (10.0)
CF_3CH_2OH (12.4)	*p*-nitrophenol (7.2)	$CH_3CH_2NH_2$ (ca40)
CH_3CONH_2 (15)	nH_3 (33)	H_2O (7)

In what way do the values of K_a and pK_a relate to the strength of an acid?

Answer

With K_a, the *higher* the value, the stronger the acid.
With pK_a, the *lower* the value, the stronger the acid.

From these figures, we can see that amides and amines are extremely weak acids (high pK_a values) compared to carboxylic acids. This can be explained by the different electronegativities of oxygen and nitrogen.

Compare now the pK_a values of ethanoic acid, ethanol and phenol. What is the order of acidity for these molecules?

Answer

Ethanoic acid is more acidic than phenol and phenol is more acidic than ethanol.

The difference in acidity is quite marked between these three structures, yet hydrogen is attached to oxygen in all three cases. There must be other factors at work apart from electronegativity. There are two important influences on acid strength other than electronegativity and these are:

- inductive effects and

- resonance effects.

4.6 Inductive effects on acid strength

We saw above that strong acids have a proton attached to a strongly electronegative atom. There are two reasons for this. First of all, the electronegative atom makes the hydrogen strongly electrophilic and more likely to react with a base. Secondly, the negative charge which results when the proton is lost is better tolerated by an electronegative atom. Strongly electronegative atoms (e.g. halogens) can tolerate the negative charge better than weakly electronegative atoms (e.g. nitrogen) and stabilise it better.

Stabilising the negative charge is therefore very important and any other mechanism which stabilises the charge will further aid the acid/base reaction and result in a stronger acid.

As stated above, there are other two other ways of stabilising a negative charge apart from electronegativity. We are going to look first at **inductive effects**.

Compare the pK_a values of the following alcohols: CF_3CH_2OH (pK_a=12.4) and CH_3CH_2OH (pK_a=16). Which is more acidic?

Answer

CF_3CH_2OH is more acidic than ethanol since it has a lower pK_a.

Clearly the difference in acidity cannot be put down to the electronegativity of the atom bearing the charge since it is oxygen in both cases. There must be another influence at work. Draw the structures of the anions which are formed when these two alcohols lose their acidic proton.

Answer

$$CH_2-\overset{..}{\underset{..}{O}}:^{\ominus}$$

F$_3$C

$$CH_2-\overset{..}{\underset{..}{O}}:^{\ominus}$$

H$_3$C

If we can explain why one of these anions is more stable than the other, then that will explain the difference in acidities between the two alcohols. Since CF_3CH_2OH is the stronger acid, would you expect the anion which is formed from it to be more stable or less stable than the anion formed from ethanol?

Answer

The more stable the anion, the more easily it is formed and the more acidic the original alcohol. Therefore, the anion $CF_3CH_2O^-$ should be more stable than $CH_3CH_2O^-$.

Clearly, the three fluorine atoms are having an effect. As you know, fluorine atoms are strongly electronegative. What effect has this on the carbon to which they are bonded?

Answer

Each C–F bond is polarised such that the neighbouring carbon atom is electron-deficient (i.e. electrophilic).

$$\begin{array}{c} \delta- \\ F \\ \\ \\ F \end{array} \overset{\delta-}{\underset{\delta-}{\overset{F}{\diagup}}} C\delta+ \overset{..}{\underset{..}{O}}:^{\ominus} \\ CH_2-$$

If the carbon atom is electron-deficient, what effect will that have on the neighbouring C–C bond?

Answer

Since the carbon is electron-deficient, it will have an electron-withdrawing effect on the electrons in the C–C bond and make the neighbouring carbon slightly electron-deficient.

This inductive effect will continue to be felt through the next bond to the oxygen atom. The effect decreases through the bonds but it is still significant enough to be felt at the negatively charged oxygen. What effect would such an inductive effect have on the charge and the stability of the ion?

Answer

The inductive effect is electron-withdrawing. This will decrease the negative charge on the oxygen and help to stabilise it. Since the anion is stabilised, it is more easily formed. This means that the original alcohol will lose its proton more readily and be a stronger acid.

Compare the acid strengths of ethylamine (pK_a ca 40) and ammonia (pK_a ca 33). Which is the weaker acid?

Answer

Ethylamine is the weaker acid because it has a higher pK_a.

Draw the anions which are formed from these molecules when they behave as acids and explain in terms of inductive effects why ethylamine is the weaker acid. (Remember that alkyl groups have electron-donating properties.)

Answer

The anion formed from ethylamine is less stable than the anion formed from ammonia due to the inductive effect of the alkyl group. Alkyl groups are electron donating and push electrons towards the nitrogen. This enhances the negative charge and destabilises the ion. This makes ethylamine a much weaker acid as a result.

4.7 Resonance effects on acid strength

We are now going to compare the acidities of ethanoic acid, phenol and ethanol. As we saw earlier, ethanoic acid is a stronger acid than phenol, and phenol is a stronger acid than ethanol. This can be seen from the strength of bases required to remove the protons from these compounds.

For example, ethanoic acid will form salts by reaction with weak bases (e.g. the bicarbonate ion) as well as with strong bases (e.g. the hydroxide ion).

Phenol, on the other hand, will only form a salt on reaction with the strongly basic hydroxide ion. It does not react with the weaker bicarbonate ion.

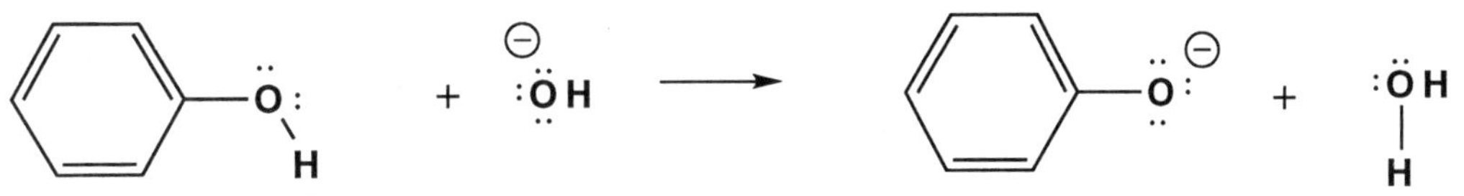

Ethanol is such a weak acid that it reacts with neither of these bases and requires treatment with sodium hydride to lose its acidic proton.

We now have to ask why these different functional groups have different acidic strengths.

Once again we will look at the anions which are formed and see how they can be stabilised. Based on the experimental facts above, decide which of ethanoate, phenolate or ethoxide is the most stable anion and which is the least stable anion.

Answer

Since the carboxylic acid is the strongest acid, the ethanoate ion must be the most stable ion. Similarly, the phenolate ion must be more stable than the ethoxide ion.

Therefore, the ethanoate ion is more stable than the phenolate ion, and the phenolate ion is more stable than the ethoxide ion.

The more stable the salt, the more likely it is to be formed and the more likely the acid/base reaction will take place. What makes the ethanoate ion more stable than the ethoxide ion? We have already seen that inductive effects can stabilise or destabilise negative charges. There is another effect which alters the stability of ions and this is called the **resonance effect**. Resonance can stabilise charge by delocalising the negative charge.

We shall look at ethanoic acid more closely.

First of all, draw a mechanism for the following reaction.

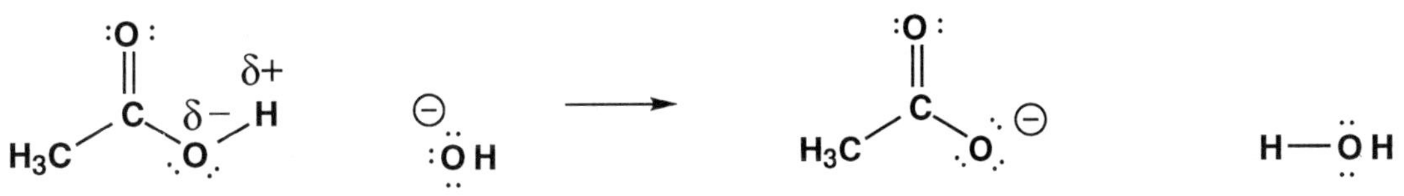

Answer

The hydroxide ion is negatively charged and is the nucleophile. The acidic proton on the carboxylic acid is electrophilic because it is bound to the more electronegative oxygen.

The electrons for the new HO–H bond are provided by the oxygen of the hydroxide ion. As the new bond is formed, the O–H bond of the carboxylic acid must break since hydrogen is only allowed one bond. Both electrons move onto the oxygen and a negatively charged carboxylate ion is formed.

Let us look at the carboxylate ion. The charge is on oxygen which is electronegative and able to stabilise that charge. However, there is another oxygen atom present in the molecule—the carbonyl oxygen. If the charge could be shared between both these oxygens, that would help to stabilise the charge even more. This is what is known as delocalisation. In other words, the charge is no longer concentrated on one atom but is shared over two or more atoms. The way in which delocalisation can take place is through resonance.

In order to get resonance, it is essential to have a π bond next door to the atom bearing the charge, i.e.

Is such a system present in the carboxylate ion? Highlight it if it is.

Answer

There *is* such a system present in the carboxylate ion. The carbonyl group contains a π bond and the negatively charged oxygen represents X.

We shall now look·at the electron movements involved in resonance, remembering that we want the negative charge to eventually end up on the other oxygen. Resonance involves an interaction between the π electrons of the carbonyl group and the lone pair of electrons responsible for the negative charge on oxygen.

In this process, new bonds are formed and old bonds are broken. Whenever bonds are formed or broken, electrons are involved and therefore we need to identify which are the most likely electrons to be used. Remembering that nucleophilic centres are the centres most likely to provide electrons for new bonds, identify the most likely pair of electrons which could be used to make a new bond.

Answer

The carboxylate ion is nucleophilic because it is negatively charged and electron rich. The nucleophilic centre is the negatively charged oxygen. This atom has three lone pairs of electrons and one of these pairs of electrons could be used to form a new bond.

Nucleophilic centre

In resonance mechanisms, the new bond must be formed within the same molecule and not to another molecule. Where then will the new bond be formed?

Answer

The new bond must be formed to the neighbouring carbon atom.

Neighbouring
carbon atom

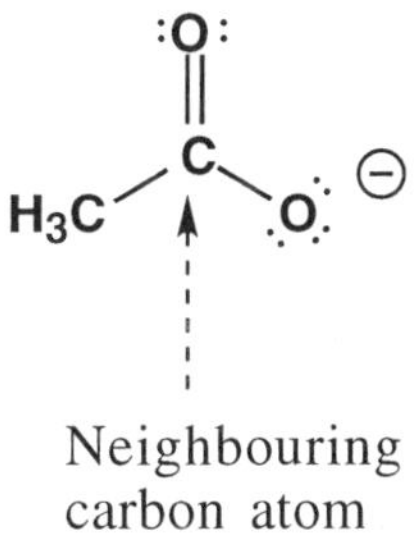

If that bond is to form what kind of bond must it be (σ or π?) and how many bonds would carbon have if it gained this new bond.

Answer

The new bond would have to be a π bond since there already is a σ bond between the carbon and the oxygen. The carbon atom would end up with five bonds and this is not allowed.

Clearly, one of the bonds already present to carbon has to break if the new bond is to form. Which is the most likely bond to break and why?

Answer

The most likely bond to break will be the π bond already present since this is the weakest bond.

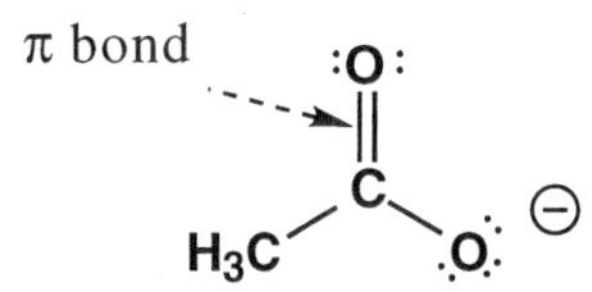

π bond

:O:

H₃C—C—:O:⁻

If the existing π bond breaks, the electrons have to go somewhere. Where must they go and why?

Answer

The only place they can go is to the electronegative oxygen to form a third lone pair of electrons. As a result, that oxygen gains a negative charge.

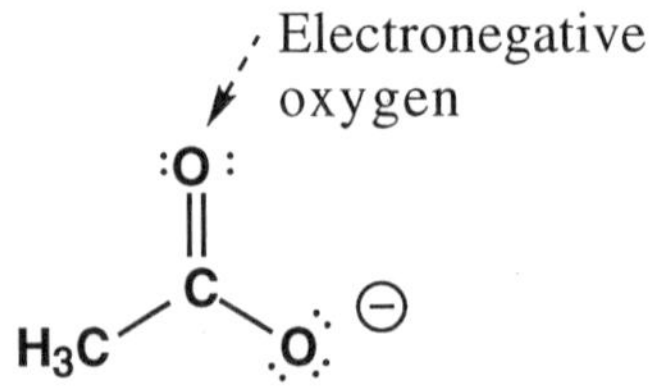

Electronegative oxygen

:O:

H₃C—C—:O:⁻

Draw the mechanism for this process.

Answer

The arrows show the lone pair of electrons on the 'bottom' oxygen forming a new π bond to the neighbouring carbon. At the same time, the weak π bond which is already present must break. Both electrons in that bond end up on the 'top' oxygen. This means that this oxygen now has three lone pairs and has the negative charge.

  Note that the π bond and the charge have effectively 'swapped places'.

  Notice also the double-headed arrow which is used between resonance structures. It means that there is resonance between these two structures and that the negative charge is shared equally between both oxygens.

Can you now see why it is necessary to have the π bond next to the negatively charged atom if resonance is to take place? Would it be possible to delocalise the charge in the following anion, for example?

Answer

Resonance is not possible here because there is no neighbouring π bond. If a lone pair of electrons on the negatively charged oxygen made a new bond to the neighbouring carbon atom, a σ bond would have to be broken and a different molecule would be obtained.

cont.

This mechanism does occur (Section 9), but it is not a resonance mechanism.

We have seen from the resonance mechanism how the charge is moved from one oxygen to the other through the movement of two sets of electrons. This can often create confusion and some typical errors are now shown. The following mechanism was suggested by one student to show how the negative charge could be delocalised between both oxygens. What is wrong with it?

Answer

The student knew the charge had to be moved to the other oxygen but could not work out how to do it. Instead, he (or she) thought that drawing an arrow to show the charge moving over would be sufficient. This is wrong! First of all, arrows show the movement of *electrons*, not *charge*. Secondly, the structure shown on the right makes no sense at all. Why not?

Answer

None of the lone pairs or bonds have been altered between the two structures, so it would be impossible for one oxygen to lose its charge and the other oxygen to gain a charge if they still have the same number of electrons.

Here is another incorrect answer. This time the student has realised that electrons have to move from one oxygen to the other if the charge is to move. What is wrong with this mechanism?

Answer

In this mechanism, the electrons have magically jumped from one atom to another. This is impossible. A lone pair of electrons cannot 'jump' from one atom to another. Furthermore, the mechanism does not explain how the π bonds have changed places.

We have looked at how the charge in a carboxylate ion can be delocalised or 'spread'. Such delocalisation helps to stabilise charge. Can you now explain why ethanoic acid is a much stronger acid than ethanol?

Answer

Ethanoic acid is a stronger acid because the carboxylate ion which is formed can be stabilised by resonance, thus delocalising the charge.

Ethanol, on the other hand, cannot delocalise its charge because resonance is not possible. Furthermore, the inductive donating effect of the alkyl group enhances the charge and destabilises it.

Let us now look at phenol and explain why phenol is less acidic than ethanoic acid but more acidic than ethanol.

Show the mechanism for the reaction of phenol with the hydroxide ion.

Answer

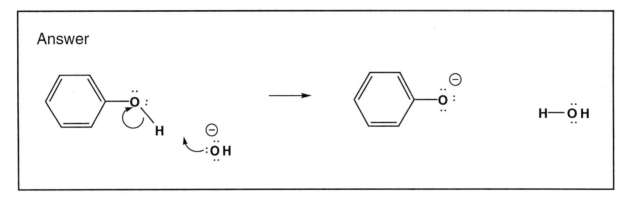

The phenolate ion must be more stable than the ethoxide ion. Is resonance possible for the phenolate ion? Remember there has to be a π bond next to the atom with the negative charge. Highlight any such system.

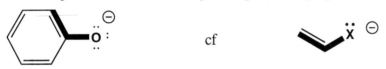

Answer

Resonance is possible because of the neighbouring π bond highlighted below.

Look again at the resonance mechanism for the carboxylate ion, then show the resonance mechanism for this system, showing where the negative charge ends up.

Answer

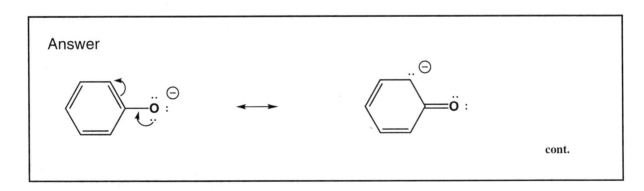

cont.

The lone pair on oxygen can form a new π bond to the neighbouring carbon on the aromatic ring. It can do so because there is a weak π bond which can be broken to allow this new bond. As the old π bond breaks, its electrons move onto the next carbon atom, giving it a negative charge.

Are these resonance structures equally likely, or is one more stable than the other?

Answer

In the first resonance structure, the charge is on an electronegative oxygen whereas on the second resonance structure, the charge is on carbon. This is not so stable and so the predominant resonance structure is the one where the charge is on the oxygen. In other words, most of the charge is on oxygen.

Look at the second resonance structure you have drawn, where the charge is on carbon. Is there any other resonance which is possible for that structure? Look again for a π bond next to the charge bearing atom.

Answer

There are two π bonds next to the charge-bearing atom in this structure.

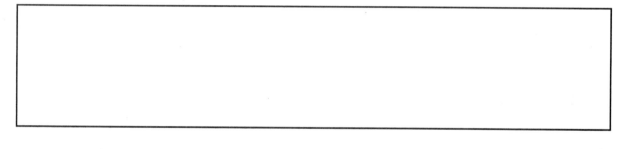

Draw the resonance mechanism for both these structures and identify any new resonance structures.

cont.

Answer

This resonance gives back the original resonance structure.

This resonance gives a new resonance structure.

Now consider the new resonance structure and work out a mechanism by which you would get yet another resonance structure.

Answer

You have now seen all the possible resonance structures for the phenolate ion. Identify the different positions where the negative charge can be delocalised on this structure.

Answer

The negative charge can be delocalised at the positions indicated.

Show in one mechanistic sequence how the four resonance structures are formed for the phenolate ion.

Answer

In the phenolate anion, the charge can be spread around three atoms of the aromatic ring. This might suggest that the phenolate anion should be more stable than the carboxylate anion since the charge is spread over more atoms. In fact it is not. Why not?

Answer

We have already discussed this. There are certainly more resonance structures available for the phenolate ion compared to the carboxylate ion (four versus two). However, the two resonance structures for the carboxylate ion are equally likely since the charge is on electronegative oxygens.

With the phenolate ion, three of the resonance structures have the charge on a carbon atom and these are far less important than the resonance structure having the charge on oxygen. As a result, the delocalisation is weaker for the phenolate ion than for the ethanoate ion.

Nevertheless, delocalisation is still taking place and this explains why a phenolate ion is more stable than an ethoxide ion.

The ethoxide ion has its charge localised on oxygen.

$$CH_3CH_2 \!-\! \overset{..}{\underset{..}{O}}: \quad \overset{\ominus}{}$$

Ethoxide ion

There are no possible resonance structures. This makes the ethoxide ion the least stable (or most reactive) of the three anions we have studied. The situation is made worse by the fact that the ethyl group has an inductive donating effect which destabilises the charge even further.

4.8 Acid strengths of amines and amides

Amines and amides are very weak acids and need very strong bases to remove the weakly acidic proton. The pK_a values for ethanamide and ethylamine are 15 and 40, respectively. Identify which has the more acidic proton and explain why.

Ethanamide   Ethylamine

Answer

Ethanamide has the more acidic proton since it has a lower pKa. It is more acidic since the anion which is formed can be stabilised by resonance.

The anion formed from ethylamine has a localised charge on the nitrogen which is further destabilised by the inductive effect of the alkyl group attached.

4.9 Lewis acids

Lewis acids are electron deficient molecules which are classed as acids because they are capable of accepting a pair of electrons from another molecule to fill their valence shell. A Lewis base is a molecule which can provide those electrons. Two common Lewis acids are BF_3 and $AlCl_3$.

Look up Al and B in the periodic table. How many valence electrons do these atoms have?

Answer

They both have three valence electrons.

How many electrons are in the valence shell of boron in BF_3?

Answer

Each bond has two electrons. Therefore, there are six electrons surrounding boron in its valence shell.

There are six electrons surrounding boron in BF_3 but the valence shell can accomodate eight and so a fourth bond is possible. Boron cannot provide any more electrons for this new bond and so both electrons have to be provided by an incoming ion to make a fourth bond.

Example

$$F^{\ominus} \quad + \quad BF_3 \quad \longrightarrow \quad F-\overset{\ominus}{BF_3}$$

Show the mechanism for this reaction.

Answer

In this reaction, BF_3 is defined as a Lewis acid because it is accepting an electron pair in the formation of the new bond.

An example of $AlCl_3$ acting as a Lewis acid is the following.

Summary

- A Brønsted–Lowry acid is a molecule which donates a proton.

- A Brønsted–Lowry base is a molecule which accepts a proton.

- Acidic protons are attached to electronegative atoms.

- Acidity is related to the electronegative strength and size of the neighbouring atom.

- Hydrogen attached to a halogen is more acidic than a hydrogen attached to oxygen. A hydrogen attached to oxygen is more acidic than a hydrogen attached to nitrogen.

- Hydrogens attached to a non-activated carbon are not acidic (for examples of an activated carbon, see Section 11).

- Acidity is related to the stability of the anion formed on loss of the proton.

- A negative charge can be stabilised by inductive or resonance effects.

SECTION 5

Electrophilic Addition

Electrophilic Addition

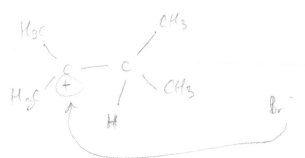

5.1 Electrophilic addition of HBr to symmetrical alkenes

One example of electrophilic addition is the reaction of an alkene with HBr.

Why is this reaction considered a functional group transformation?

(empty box)

Answer

The reaction has transformed an alkene functional group into an alkyl halide functional group.

Let us now look at this reaction in some detail.
Identify the nucleophilic and electrophilic centres in each reagent.

Answer

H₃C, CH₃
   \   /
    C=C
   /   \
H₃C   CH₃

Nucleophilic
centre

$\delta +$ .. $\delta -$
H — Br:
      ..

Electrophilic
centre

(Note: Bromine is a nucleophilic centre, but it is such a weak nucleophile that we can usually ignore it.)

The alkene is a nucleophile while the HBr is an electrophile. Reaction should lead to a new bond between these molecules. Where do the electrons come from for this new bond?

Answer

The electrons for the new bond come from the nucleophile and specifically from the nucleophilic centre (i.e. the alkene double bond).

There are two bonds in the double bond. One is a σ bond and the other is a π bond. One of these bonds has to break to provide the two electrons for the new bond. Which do you think is the most likely bond to break?

Answer

The π bond is the one which will break to provide the two electrons for the new bond. This is because the π bond is weaker than the σ bond and more easily broken. The π molecular orbital is formed by the 'side on' overlap of p atomic orbitals. There is less overlap compared to the strong 'head on' overlap of the sp² orbitals in the σ bond.

We have established that the electrons for the new bond are provided by the alkene. Where will the new bond go to on the HBr molecule?

Answer

The bond will be formed to the electrophilic centre (i.e. the H).

If a new bond is formed to the H of HBr what also has to happen? (Remember hydrogen is only allowed one bond.)

Answer

The bond between hydrogen and bromine has to break, with both electrons ending up on the bromine. We can represent this with the following arrows.

$$\text{Alkene:} \quad \overset{\delta+}{\text{H}}\text{—}\overset{\delta-}{\text{Br}}: \quad \longrightarrow \quad \left[\text{Alkene–H}\right]^{\oplus} + \quad :\overset{\ominus}{\text{Br}}:$$

Note that the arrow from the alkene is pointing to where the centre of the new bond will be located and not to the hydrogen itself. Remember that a mechanism arrow only points to an atom if the electrons involved are going to form a lone pair on that atom.

We shall now look at what happens in more detail. The alkene π bond is used for the new bond and we draw the following arrows to show the mechanism. (Note again how the arrow from the alkene points to where the centre of the new bond will be formed.) Draw the structures which are obtained.

$$\begin{array}{c}
\text{H}_3\text{C} \qquad\qquad \text{CH}_3 \\
\diagdown \diagup \\
\text{C}\!=\!\text{C} \\
\diagup \diagdown \\
\text{H}_3\text{C} \qquad\qquad \text{CH}_3
\end{array}$$

$$\begin{array}{c}
\text{H} \quad \delta+ \\
| \\
:\text{Br}: \quad \delta-
\end{array}$$

Answer

The structures obtained are as follows.

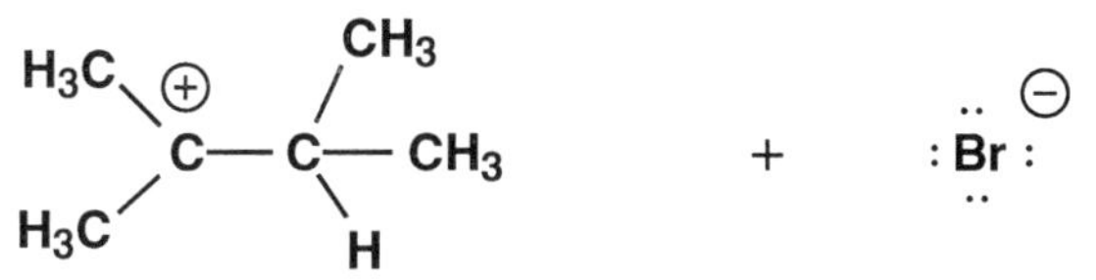

H₃C, ⊕ ... CH₃ ... C—C—CH₃ ... H₃C ... H   +   :Br:⊖

Note that the two electrons of the original π bond have 'swivelled' round the 'right-hand' alkene carbon to form a σ bond to hydrogen. Consequently, the other carbon has only three bonds and is positively charged. This is a reaction intermediate.

The positively charged intermediate is known as a **carbocation** since the positive charge is on the carbon atom. This is a reactive species and will not survive for very long with the bromide ion in the vicinity. A reaction will occur between the bromide ion and the carbocation. Can you work out logically what that reaction will be?

If you are unsure, ask yourself the following questions. Which ion is electrophilic and which one is nucleophilic? Where are the electrophilic or nucleophilic centres? Where will a new bond be formed? Where will the electrons for the new bond come from? Where will the new bond go to? How would you draw arrows to show what is happening? What product will be obtained?

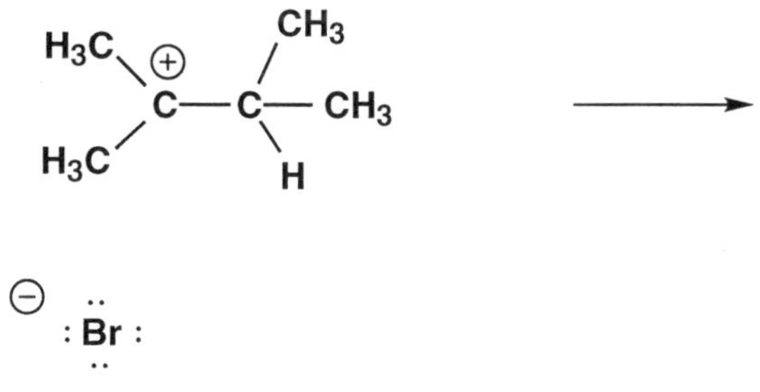

H₃C, ⊕ ... CH₃ ... C—C—CH₃ ... H₃C ... H   ⟶

⊖ :Br:

Answer

Let us go through the list of questions given above. The carbocation is an electrophile since it is positively charged. The bromide ion is a nucleophile since it is negatively charged. The electrophilic centre on the carbocation is the carbon which is positively charged. The bromide ion is obviously the nucleophilic centre. A nucleophile will react with an electrophile with the new bond formed between the nucleophilic centre and the electrophilic centre. The nucleophile always supplies the electrons and therefore the electrons for the new bond will come from one of the bromide's lone pairs.

cont.

The bond will go to the carbon bearing the positive charge. The mechanism to show this is as follows, and this gives the final product.

We have now gone through the whole mechanism in detail. Write down the complete mechanism and explain why the process is called electrophilic addition.

Answer

The complete mechanism is as follows.

The mechanism is known as an addition reaction since the overall reaction involves the addition of HBr to the alkene. It is an electrophilic addition since the first step of the mechanism involves the addition of the electrophilic H to the alkene. Note that the second step involves a nucleophilic addition to the carbocation intermediate, but it is the first step which defines the reaction.

Let us return to the first stage of the mechanism. As you will recall, the π electrons of the alkene provide the electrons for the new bond by swivelling round one of the carbons making up the C=C. We chose to swivel round the 'right-hand' carbon. Is there any reason why we couldn't have swivelled around the 'left-hand' carbon? What would have been the result if we had? Draw the mechanism for this and identify the products which are obtained.

Answer

The mechanism would be as follows.

If you compare the carbocation and the product above with those from the first mechanism, you should see that they are the same. Therefore, in this example it does not matter which end of the double bond is used for the new bond to hydrogen. The end product is the same. In reality, the chances are equal of the hydrogen adding to one side or the other.

Predict the product which would be obtained if the following alkene was treated with HBr and give the mechanism for the reaction.

(*E*)-2-Butene

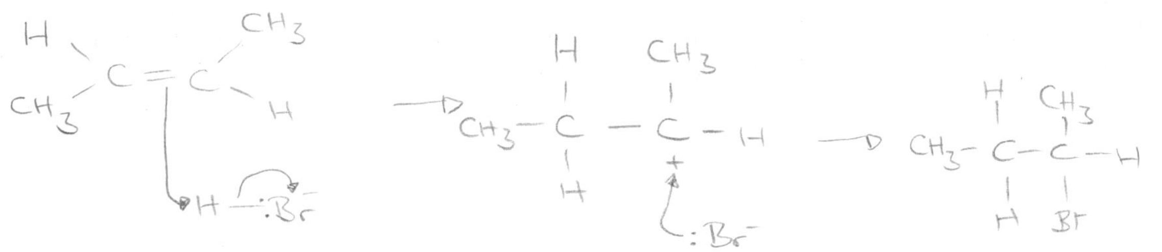

Answer

The product and mechanism are shown below. The same product would be obtained if the hydrogen added on to the 'right-hand' carbon instead.

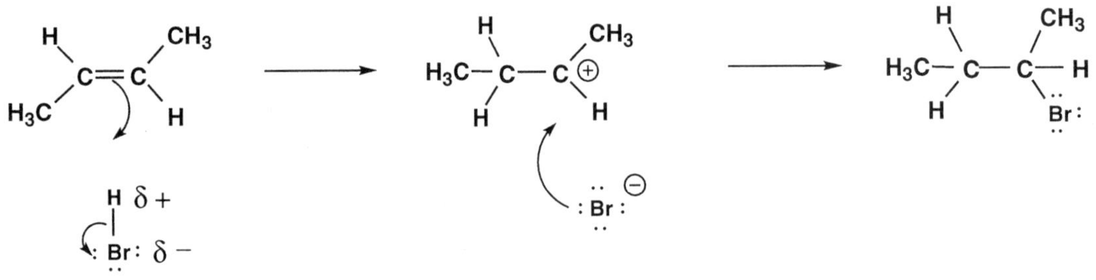

Predict the product which would be obtained if the (Z)-isomer of 2-butene was treated in the same way and compare this with the product obtained from the (E)-isomer.

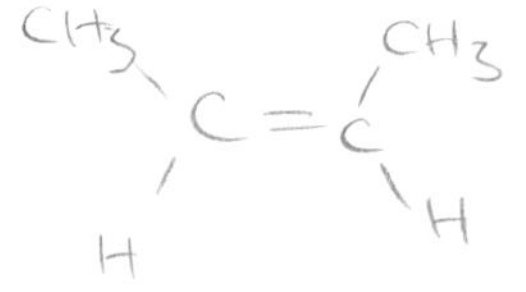

Answer

The product obtained would be the same alkyl bromide as obtained from (*E*)-2-butene. The (*Z*) and (*E*) isomers of alkenes are different compounds because there is no rotation around the double bond. In this reaction, the double bond is lost and the product has a single bond which can freely rotate. The products are therefore the same.

5.2 Electrophilic addition of HBr to unsymmetrical alkenes–Markovnikov's rule

So far, we have seen that the same product is obtained regardless of whether the hydrogen of HBr is added to one end of the double bond or the other. However, this is not always the case. Predict the products which could be obtained if the following alkene was treated with HBr.

Answer

In this case, two different products can be obtained depending on whether the hydrogen adds to one end of the alkene or the other.

cont.

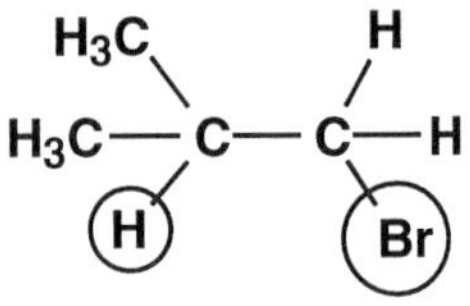

Hydrogen added to 'left-hand' end of double bond.

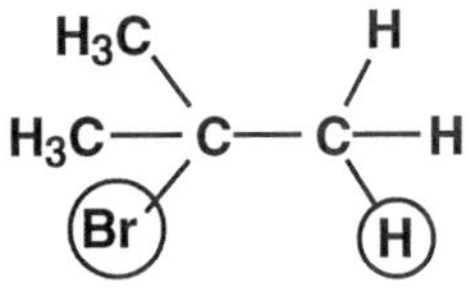

Hydrogen added to 'right-hand' end of double bond.

Why are two different products obtained in this case but not in previous cases?

Answer

In this last case the substituents at either end of the double bond are different whereas they were the same in previous cases.

Which of the following alkenes would give a single product on reaction with HBr and which would give a mixture of two products? Draw the products obtained.

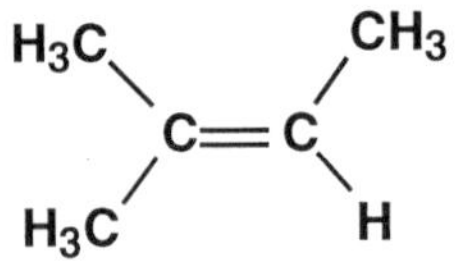

2-Methyl-2-butene

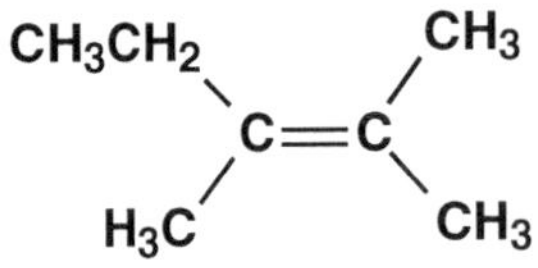

2,3-Dimethyl-2-pentene

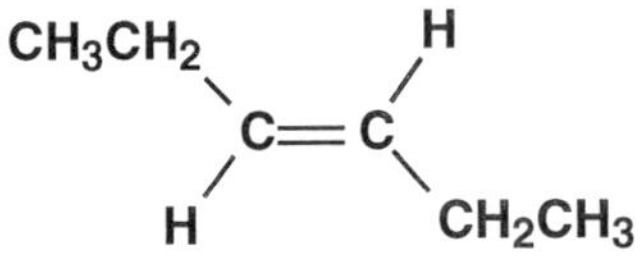

(E)-3-Hexene

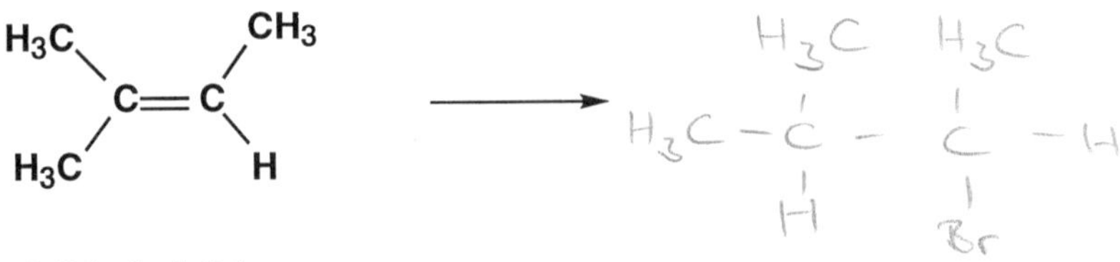

2-Methyl-2-butene

cont.

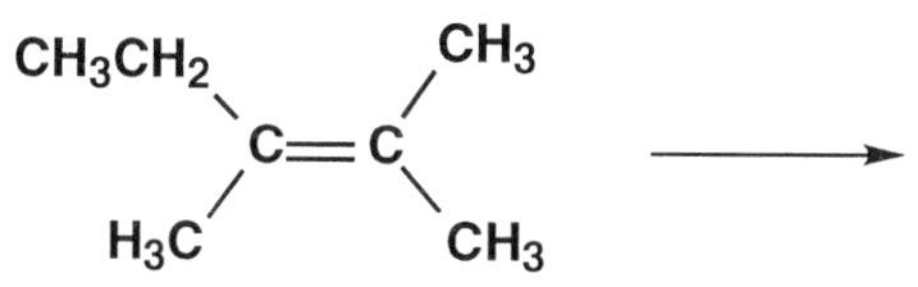

CH$_3$CH$_2$ and H$_3$C groups on one carbon; CH$_3$ and CH$_3$ on the other carbon

C=C

2,3-Dimethyl-2-pentene

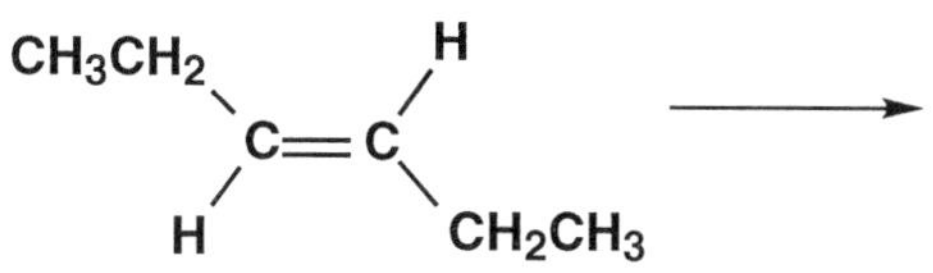

⟶

CH$_3$CH$_2$ and H on one carbon; H and CH$_2$CH$_3$ on the other

C=C

(*E*)-3-Hexene

⟶

Answer

(*E*)-3-Hexene is the only alkene above which would give a single product. Both ends of the alkene have a hydrogen atom and an ethyl group, and it does not matter which end of the alkene the hydrogen adds onto first.

The product obtained is as follows.

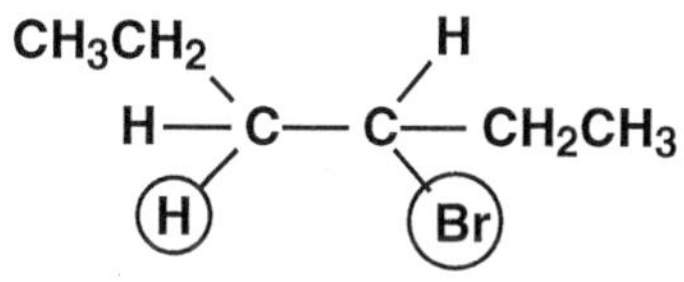

$$\text{CH}_3\text{CH}_2 \quad \text{H}$$
$$\text{H—C—C—CH}_2\text{CH}_3$$
$$\quad\text{(H)}\quad\text{(Br)}$$

The other two alkenes will give a mixture of compounds since the substituents are different at each end of the alkene.

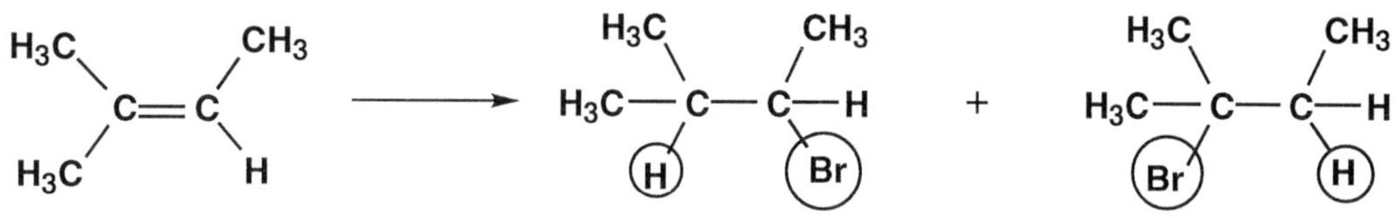

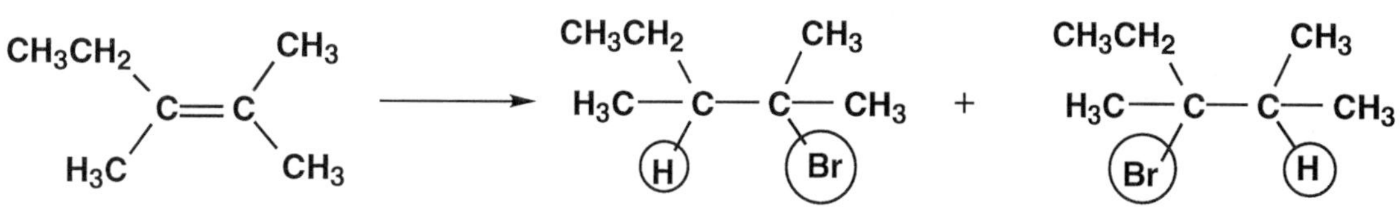

H$_3$C, CH$_3$, H$_3$C, H   C=C

⟶   H$_3$C—C—C—H   +   H$_3$C—C—C—H

(H) (Br)   (Br) (H)

CH$_3$CH$_2$, CH$_3$, H$_3$C, CH$_3$ C=C

⟶   H$_3$C—C—C—CH$_3$ +   H$_3$C—C—C—CH$_3$

(H) (Br)   (Br) (H)

Let us now return to one of the reactions which gave a mixture of products.

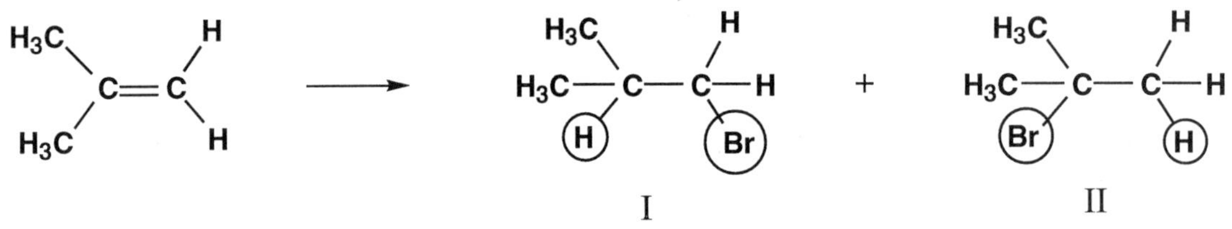

I   II

It is found that these two products are not formed to an equal extent. In fact, the major product is product II. This finding has been rationalised by proposing that the carbocation intermediate leading to product II is more stable than the carbocation intermediate leading to product I. Draw the mechanisms leading to products I and II and hence identify the more stable carbocation intermediate.

Mechanism 1

Product I

Mechanism 2

Product II

Answer

The relevant mechanisms are as follows:

cont.

Mechanism 1

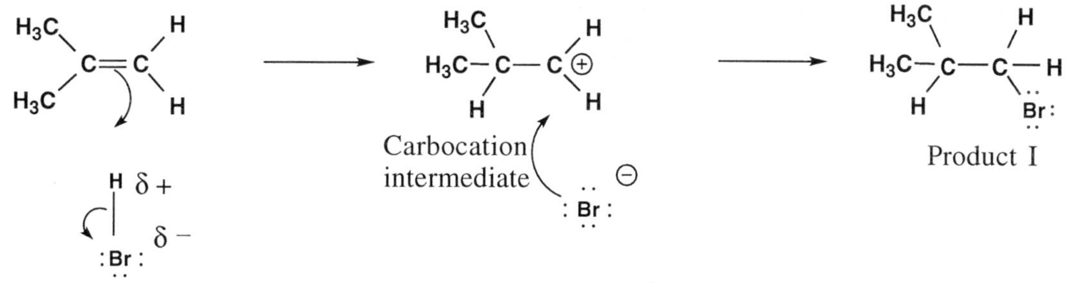

Mechanism 2

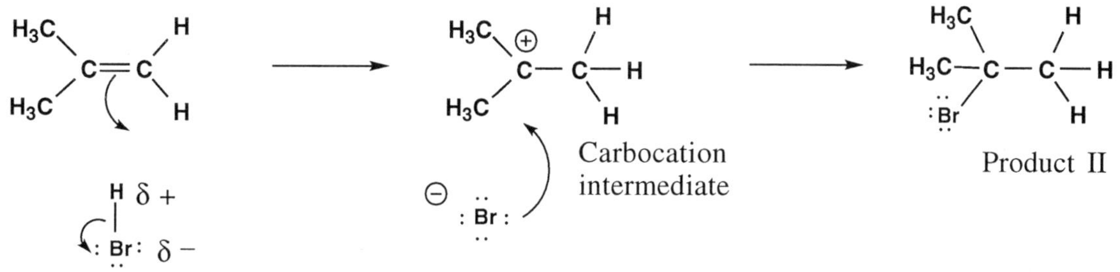

Since product II is the major product, the more stable carbocation intermediate is the one in mechanism 2.

It is easy enough to tell whether one carbocation is more stable than another if we know which product is obtained in higher yield. However, it is more useful if an organic chemist can predict in advance whether one carbocation is more stable than another and hence predict which product is more likely. Compare the two carbocations. Identify any differences you can see between them which might give some clue as to why one is more stable than the other.

Answer

If we compare the two carbocations, we can see that the more stable carbocation has three alkyl substituents attached to the positively charged carbon, whereas the less stable carbocation only has one alkyl substituent attached to the positively charged carbon.

cont.

More stable carbocation

Three alkyl groups
attached to positive
centre.

One alkyl group
attached to positive
centre.

The number of alkyl groups attached to the positively charged centre appears to have some influence on the stability of the carbocation. Why is this the case?

There are two main reasons:

- inductive effects and

- hyperconjugation.

Inductive effects

Alkyl groups are known to have an electron-donating property (i.e. they can donate or push electrons towards neighbouring groups). How would this property increase the stability of a carbocation?

Answer

Electron-donating groups such as alkyl groups push electrons towards the positively charged carbon atom and reduce the positive charge. This makes the carbocation more stable and more easily formed. The more alkyl groups which are attached, the greater the electron donating power and the more stable the carbocation.

More stable

Hyperconjugation

Before we look at hyperconjugation, we need to look at the electronic structure of the carbocation.

You already know that alkene carbons are sp^2 hybridised. What happens to the electrons in the π bond when an alkene reacts with an electrophile such as H^+? What happens to the hybridisation of the original alkene carbons when the carbocation intermediate is formed?

$$H_3C \diagdown \underset{C}{sp^2} \underset{C}{sp^2} \diagup H \qquad H^{\oplus} \longrightarrow \qquad H_3C \diagdown \overset{\oplus}{C} - \underset{H}{C} \overset{H}{\diagup} - H$$

Carbocation
intermediate

Answer

$$H_3C \diagdown \underset{C}{sp^2} \underset{C}{sp^2} \diagup H \qquad H^{\oplus} \longrightarrow \qquad H_3C \diagdown \overset{\oplus}{C} - \underset{sp^2}{C} \overset{H}{\diagup} - H$$

Both the electrons in the π bond are used to form a new σ bond to hydrogen. As a result, the carbon which gains the hydrogen becomes an sp^3 centre. The other carbon remains an sp^2 centre.

Since the carbocation has an sp^2 centre, what orbitals does the centre have and how are they used in bonding?

Answer

The carbocation centre has three sp^2 hybridised orbitals and one vacant 2p orbital. The sp^2 hybridised orbitals are used in three σ bonds. The vacant 2p orbital is not involved in bonding.

Draw the shape of the carbocation showing the position of the vacant 2p orbital.

Answer

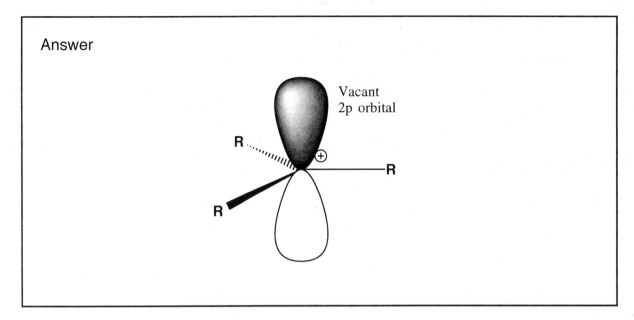

The carbocation is a reactive and unstable species due to the positive charge. We have seen already that inductive effects can help to stabilise the positive charge and thus make the carbocation more stable and hence more easily formed.

Hyperconjugation is an alternative method by which the positive charge can be stabilised. This is achieved by the overlap of the vacant 2p orbital above with a neighbouring C–H σ bond orbital. This is illustrated below.

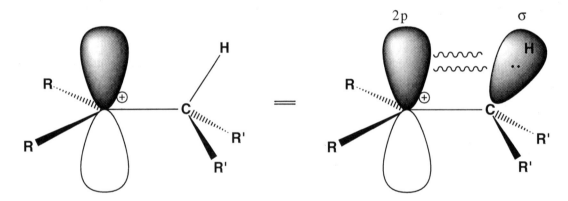

This interaction allows neighbouring σ bond electrons to stabilise the positive charge by spreading the charge out, or delocalising it. The more alkyl groups attached to the carbocation, the more possibilities there are for hyperconjugation and the more stable the carbocation.

Consider now the two possible carbocation intermediates from the earlier reaction.

Identify the σ C–H bonds which could be used to stabilise the carbocation intermediates and hence identify the more stable carbocation.

Answer

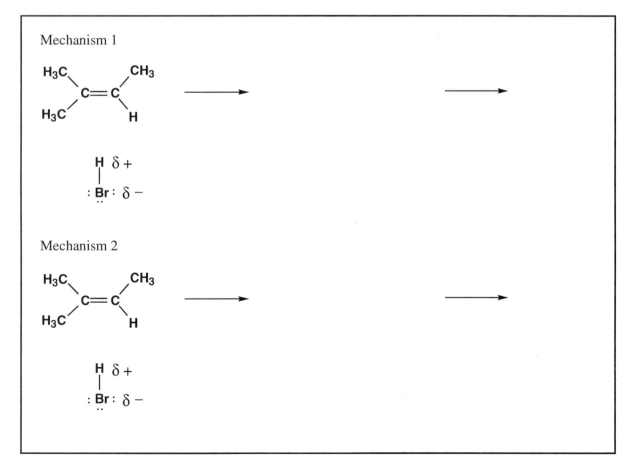

The more substituted carbocation can be stabilised by hyperconjugation to nine C–H bonds, whereas the less substituted carbocation can only be stabilised by hyperconjugation to one C–H bond. The more substituted carbocation is the more stable carbocation.

Consider now one of the alkenes which we looked at earlier.

2-Methyl-2-butene

We have already identified the two possible products which can be obtained from this alkene. Using the arguments which we have just followed, draw the possible mechanisms and predict which of the two possible products is likely to be formed in greater yield.

Mechanism 1

Mechanism 2

Answer

The relevant mechanisms are as follows:

Mechanism 1

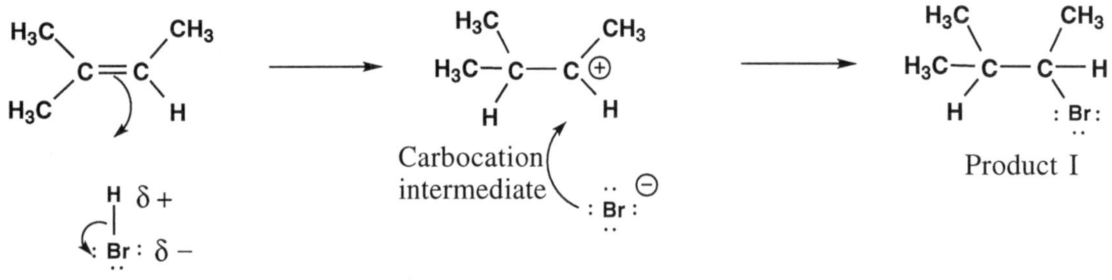

Mechanism 2

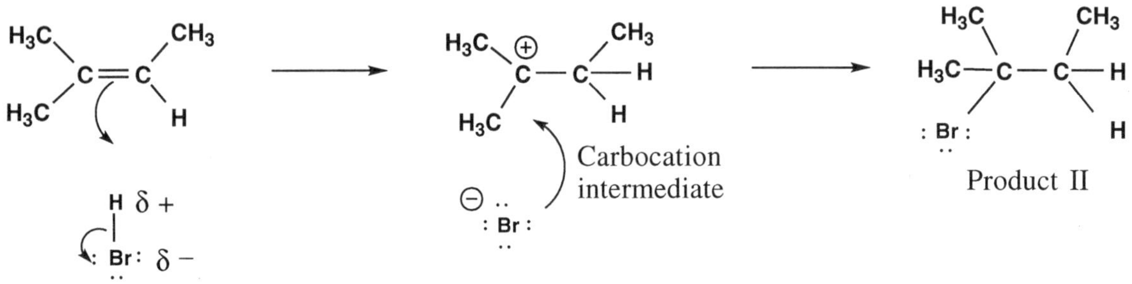

If we compare the carbocations, then we would predict the one in mechanism 2 to be the more stable since there are three alkyl groups attached to the positive centre compared to two alkyl groups in the alternative carbocation. Therefore, product II should be formed in higher yield.

Inductive effects

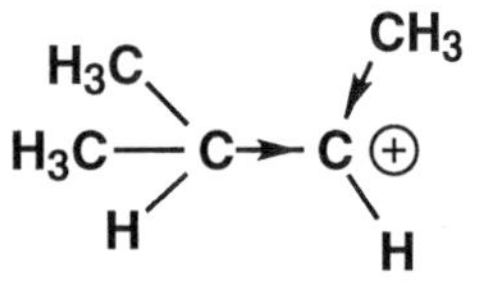

Two alkyl groups stabilising the carbocation.

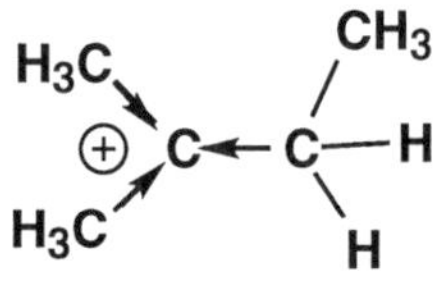

Three alkyl groups stabilising the carbocation; more stable.

Hyperconjugation

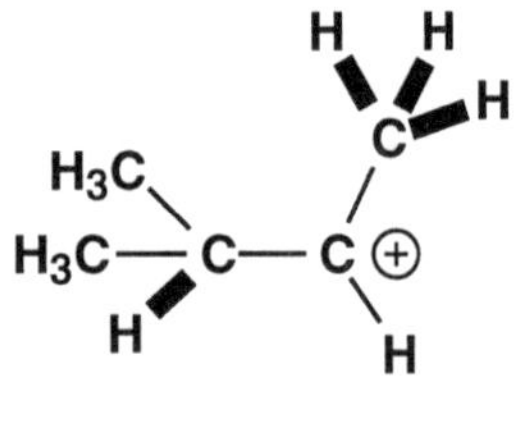

Four available C–H bonds available for hyperconjugation.

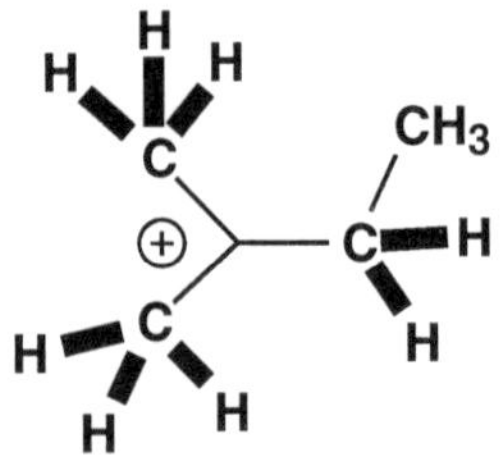

Eight C–H bonds available for hyperconjugation; more stable

Repeat the last problem with the following alkene.

$$CH_3CH_2 \quad \quad CH_3$$
$$\diagdown \quad \quad \diagup$$
$$C = C$$
$$\diagup \quad \quad \diagdown$$
$$H_3C \quad \quad CH_3$$

Answer

The relevant mechanisms are as follows:

Mechanism 1

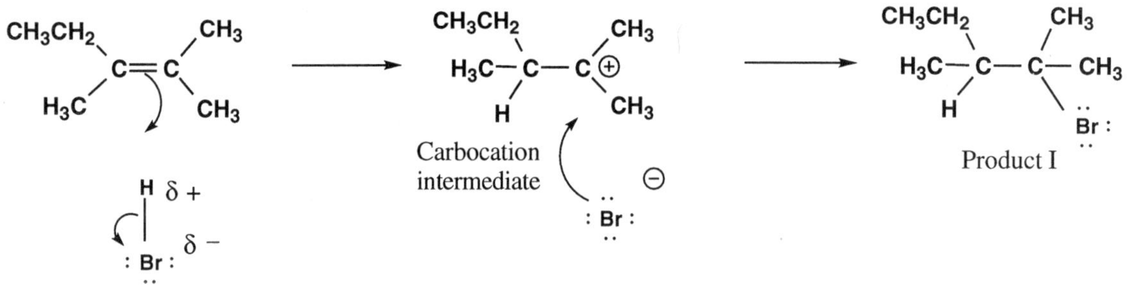

Mechanism 2

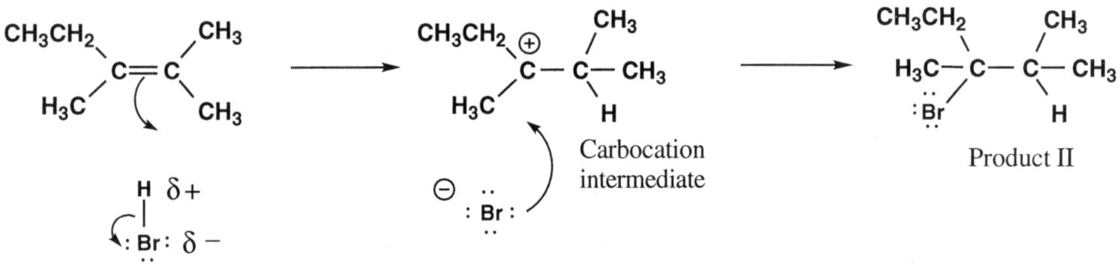

Let us now look at the carbocation intermediates in each of these mechanisms and how they can be stabilised.

Inductive effects

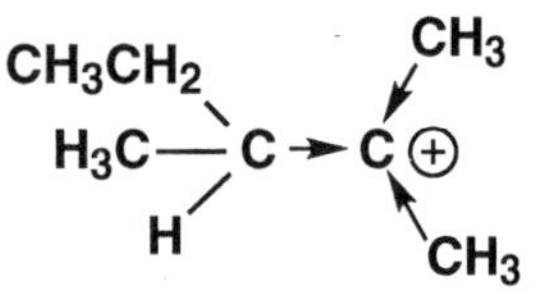

Three alkyl groups stabilising the carbocation.

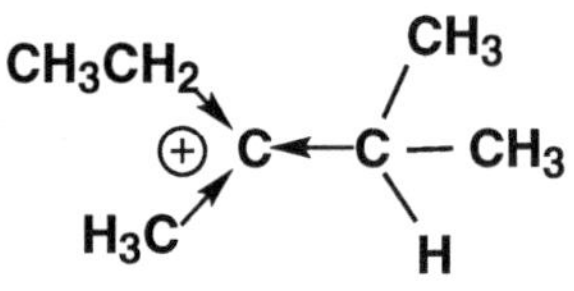

Three alkyl groups stabilising the carbocation.

Hyperconjugation

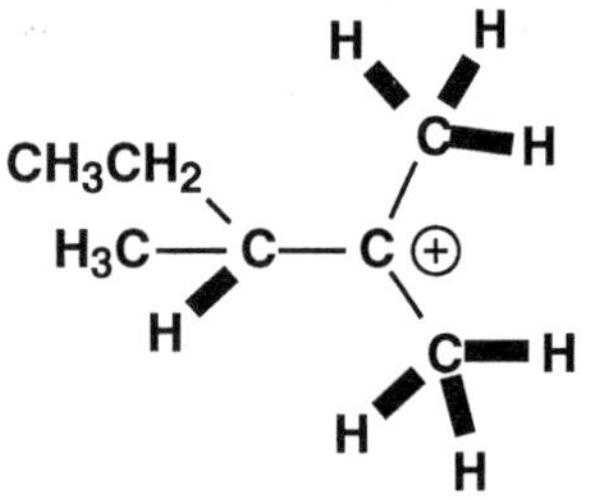

Seven C–H bonds available for hyperconjugation.

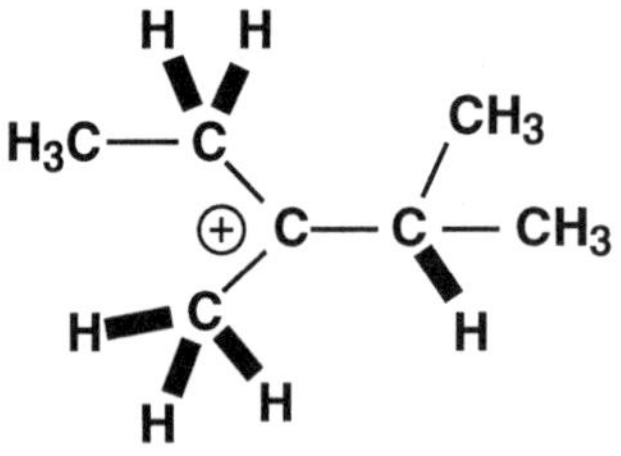

Six C–H bonds available for hyperconjugation.

Both carbocations have three alkyl groups attached to the carbocation and, as a result, the difference in stability will be small. Therefore, you might expect both products to be formed and a mixture to be obtained.

Markovnikov's rule summarises the preference with which a reagent such as HBr adds to an alkene. The rule is as follows:

> In the addition of HX to an alkene, the H attaches to the carbon with fewer alkyl substituents, and the X attaches to the carbon with more alkyl substituents.

An alternative way of stating Markovnikov's rule is to say that HX adds to an alkene such that the more highly substituted carbocation intermediate is formed in preference to the less highly substituted carbocation.

5.3 Electrophilic additions of HCl, HI and H$_2$O to alkenes

Electrophilic additions to alkenes will also take place with reagents such as HCl, HI and H$_2$O/H$^+$. The mechanism is the same. Give the products which you would expect for the following reactions.

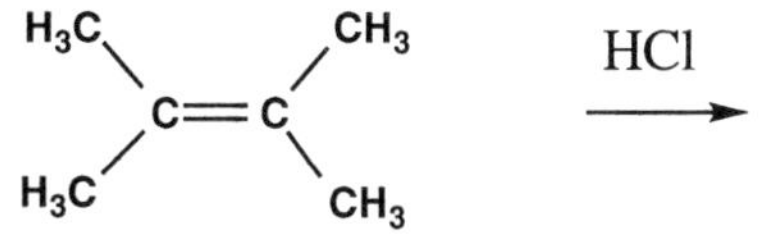

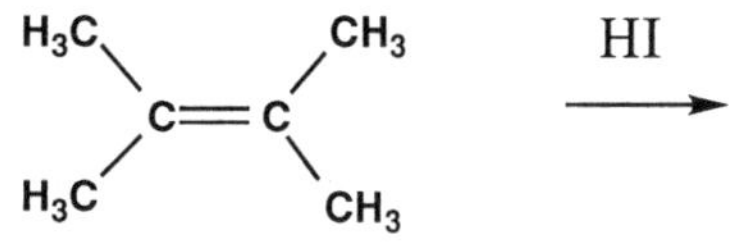

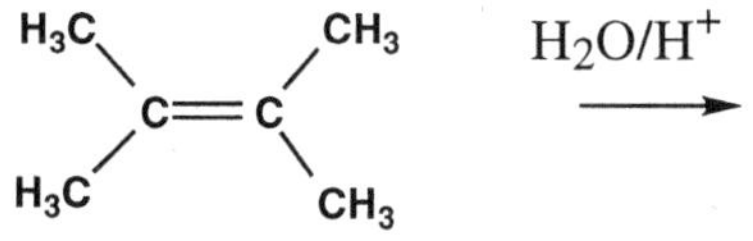

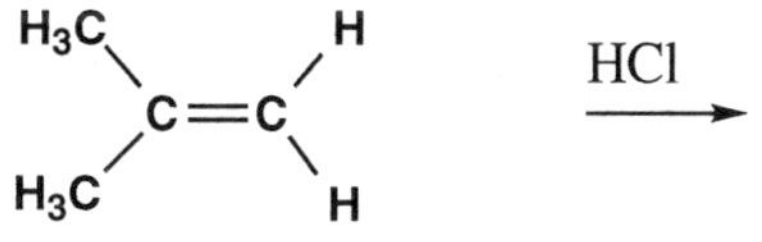

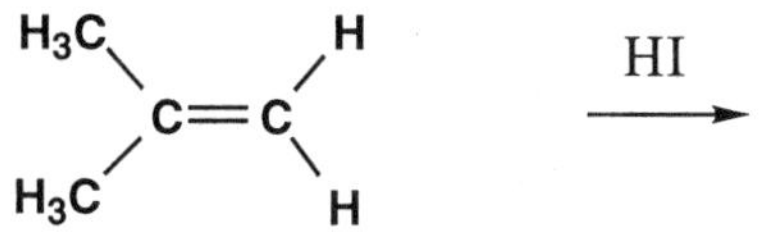

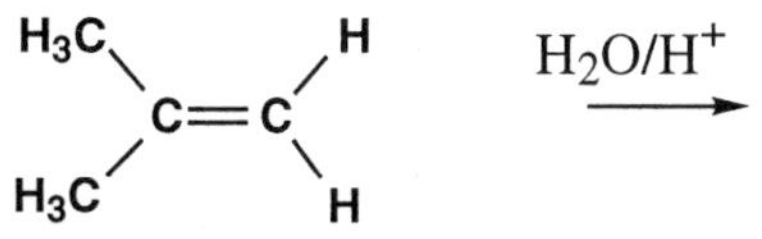

Answer

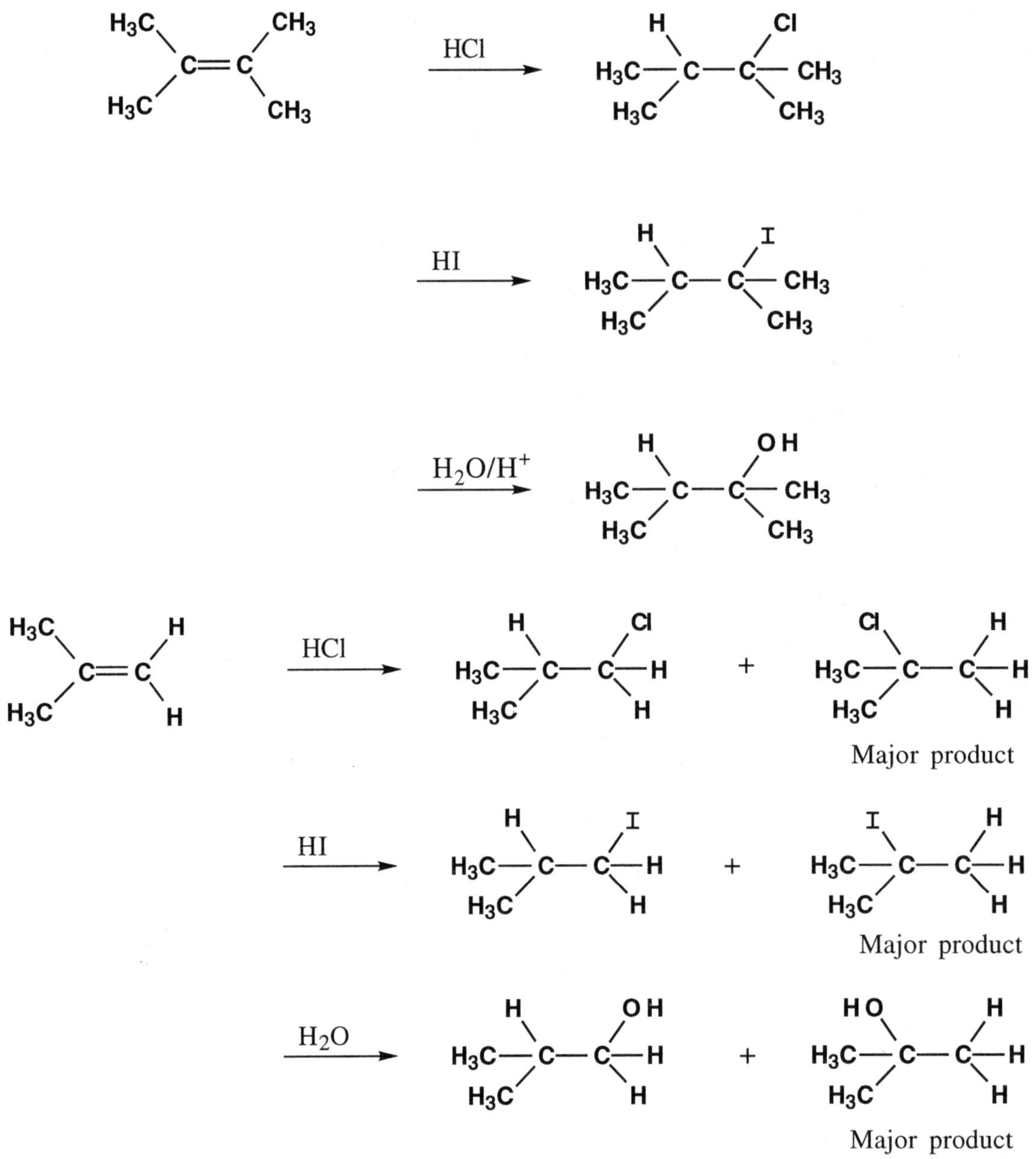

$$H_3C-C(CH_3)=C(CH_3)-CH_3 \xrightarrow{\text{HCl}}$$

5.4 Electrophilic additions of Br_2, I_2 and Cl_2 with alkenes

Electrophilic additions to alkenes are also possible with the reagents: Br_2, I_2 and Cl_2. Give the products you would expect for the following reactions.

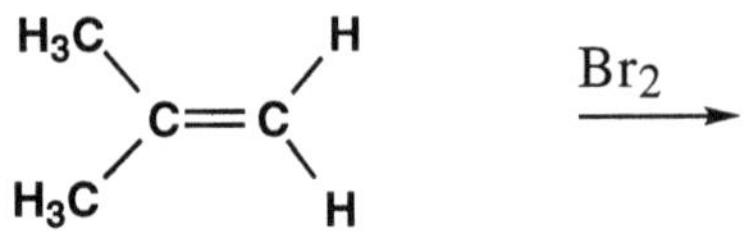

cont.

Answer

cont.

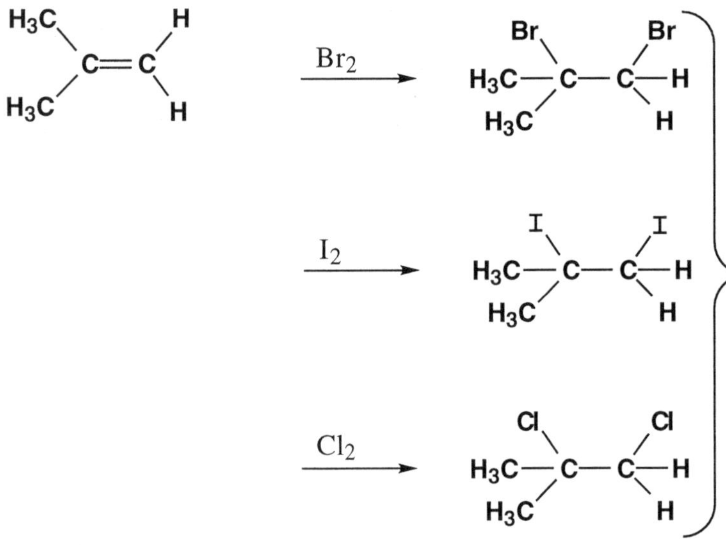

Note that the atoms added to either side of the double bond are the same. Therefore, only one product is obtained.

It is perhaps not clear why this last set of reactions are electrophilic additions since there is no obvious electrophilic centre in the reagents Br_2, I_2 or Cl_2. We need to look at the mechanism again.

First of all, describe the bonding in Br_2. Is there any electrophilic centre in the molecule?

Answer

The bond between the two bromine atoms is a covalent σ bond with both electrons equally shared between both bromine atoms. Consequently, there is no electrophilic centre. If anything, you might expect the bromine atoms to be nucleophilic because of the number of lone pairs around them.

This provides a bit of a mystery. If there is no electrophilic centre, how can the molecule react with a nucleophilic alkene? The answer lies in the fact that an electrophilic centre is induced when the Br_2 molecule approaches the alkene. The diagram below shows a bromine molecule which has come quite close to the alkene double bond. Can you suggest how an electrophilic centre can be induced in the bromine molecule?

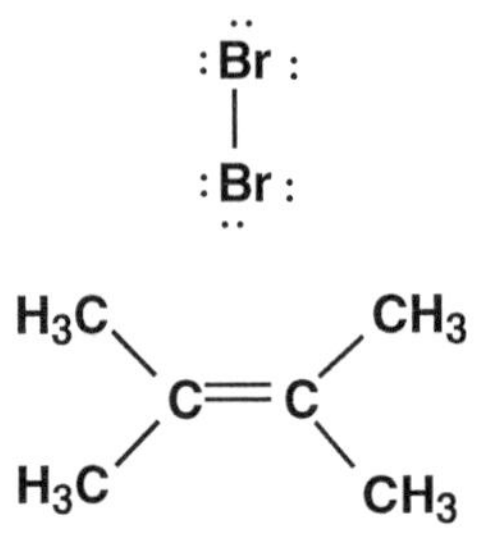

Answer

Since the alkene double bond is electron-rich it can polarise the Br–Br bond. The electrons in the double bond tend to repel the electrons present in the bromine molecule. When the bromine approaches as shown in the diagram, this repulsion will be felt more at one bromine than the other. As a result, one of the bromines becomes electron-deficient (electrophilic) while the other becomes electron-rich (nucleophilic).

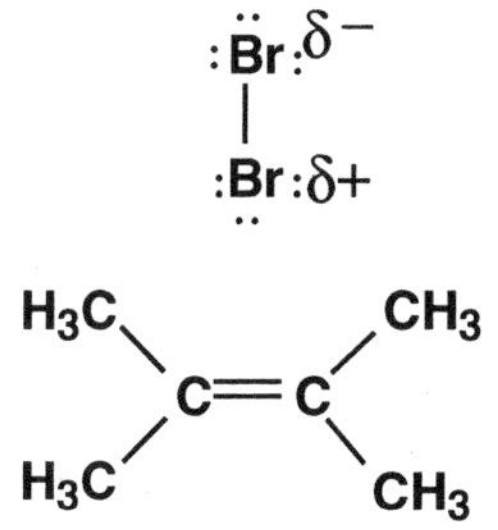

Now that an electrophilic centre has been generated in the bromine molecule, propose a mechanism leading to the product obtained.

Answer

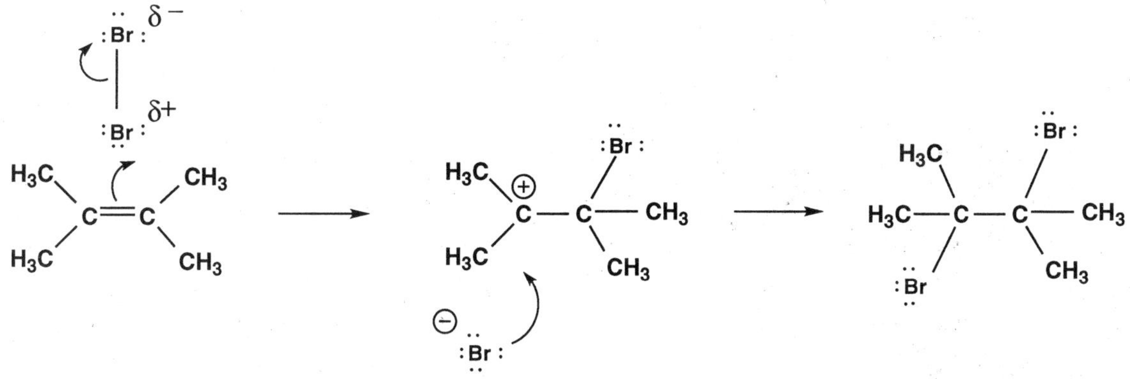

cont.

Remember to keep track of all the electrons in the mechanism. In the first step, the two π electrons of the double bond form a bond to the electrophilic Br. At the same time as this bond is being formed, the Br–Br bond is broken with both electrons moving onto the second bromine atom to give a bromide anion having four lone pairs of electrons. In the second stage, the nucleophilic bromide forms a bond to the electrophilic carbocation. Both electrons for this bond are provided by the bromide anion.

Whenever you have a charged intermediate, the reaction is more likely to proceed if the charge can be stabilised in some way. A positive charge can be stabilised if there are neighbouring alkyl groups since these electron-donating groups can 'push' electrons towards the positive centre and can also participate in hyperconjugation. As we mentioned earlier, hyperconjugation helps to stabilise the positive charge by 'spreading the charge' and 'sharing it' around different atoms in the molecule. The intermediate above can be further stabilised by sharing the positive charge between the bromine atom and a second carbon atom as shown below.

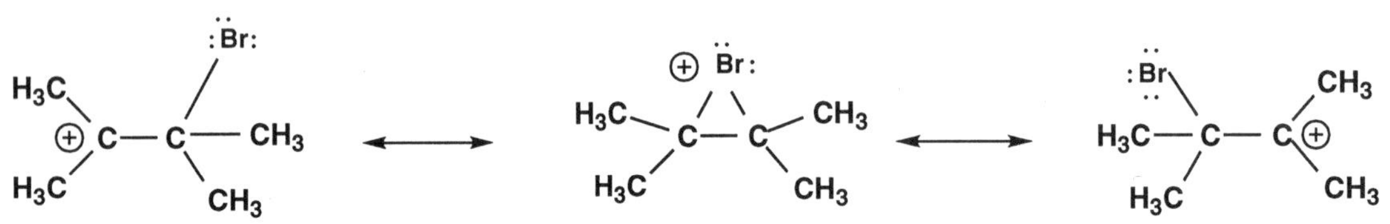

Identify any nucleophilic and/or electrophilic centres in these molecules and suggest a mechanism by which the positive charge can be spread as shown above. Identify which electrons are involved in the making or breaking of bonds.

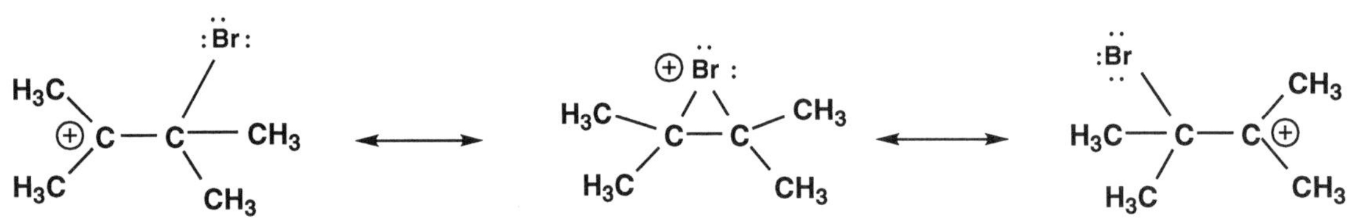

Answer

If we look at the first structure above, the positively charged carbon is an obvious electrophilic centre. The bromine has three lone pairs of electrons and is electron-rich. Therefore, it could be a nucleophilic centre. A neutral halogen does not normally act as a nucleophile despite having lone pairs of electrons. However, in this case the halogen is held close to the carbocation and is more likely to react as a result. If the lone pair of electrons on bromine is used to form a bond to the positively charged carbon, then the bromine will gain a positive charge. The structure obtained is known as a bromonium ion.

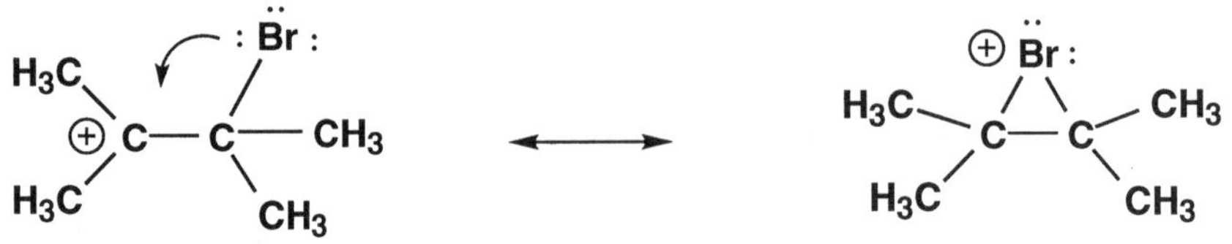

cont.

The mechanism can go in reverse to regenerate the original carbocation. Alternatively, the other carbon–bromine bond can break with both electrons moving onto the bromine. This gives a second carbocation where the other carbon bears the positive charge.

If the carbocation resulting from addition of 'Br$^+$' can be stabilised via a bromonium ion, could the carbocation resulting from addition of H$^+$ be stabilised in the same way?

In other words, is the following mechanism possible?

Answer

This mechanism is impossible. The first arrow suggests that hydrogen can provide a pair of electrons to form the three membered ring. However, hydrogen has no lone pairs of electrons with which to do this.

Test your knowledge of electrophilic additions by explaining what would happen if the following molecule was treated with bromine, and give the full mechanism for the reaction. Show how the intermediate can be stabilised.

Answer

The full mechanism for the reaction is shown below to give the product (P).

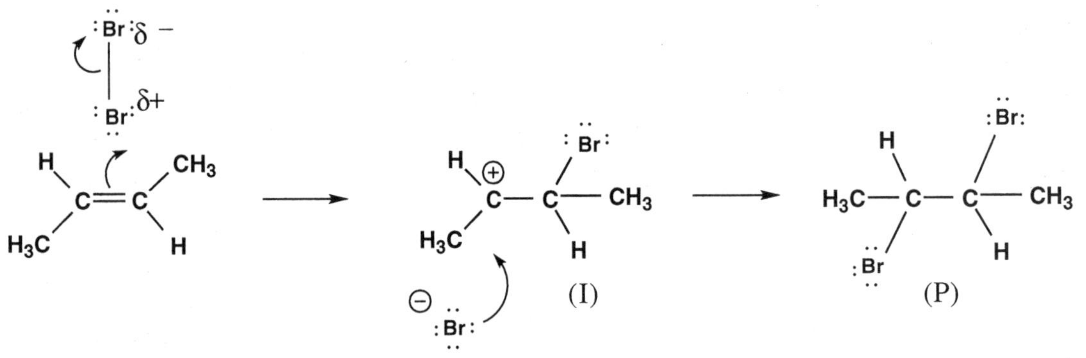

The intermediate (I) can be stabilised by the following mechanism

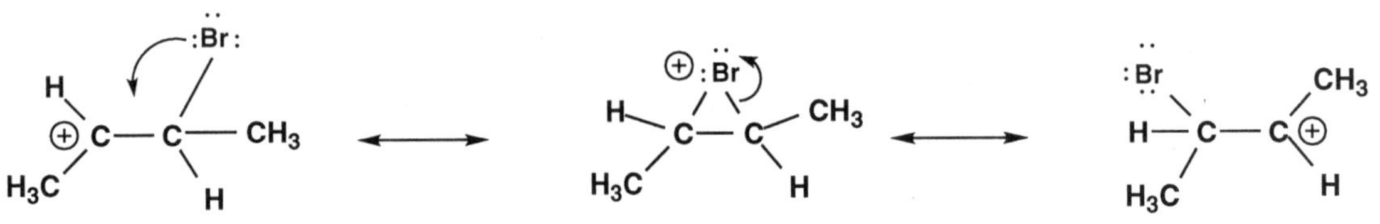

We have seen how the reaction of alkenes with bromine, iodine or chlorine gives a dihaloalkane product. However, if the reaction with bromine is carried out in water, a different product is obtained (see below).

Without going into detail, can you suggest why a bromoalcohol is obtained in the second reaction rather than the dibromoalkane. Where is the OH group coming from?

Answer

It would appear that water is taking part in the reaction to provide the OH group.

We now have to explain how this product can be obtained. To do this, we need to look at the original mechanism for the electrophilic addition of bromine.

Again without going into detail, suggest at which stage in the above mechanism water would be involved to give the bromoalcohol product.

Answer

If water replaced the bromide anion as the nucleophile at stage two of the mechanism, then the bromoalcohol would be obtained.

What is the nucleophilic centre on water?

Answer

The nucleophilic centre must be the oxygen since it is more electronegative than hydrogen, and also has two lone pairs of electrons.

Show how water could bond to the carbocation intermediate above and suggest how the structure formed would be converted to the final product.

Answer

The water uses a lone pair of electrons on the oxygen to form the bond to the carbocation. As a result, the oxygen has to share 2 electrons and gains a positive charge as a result. This charge is lost and the oxygen regains its second lone pair when one of the O–H bonds breaks and both electrons move onto the oxygen. A proton leaves the structure.

In examinations, students often draw the arrow for the final stage of the mechanism incorrectly (see below).

Both these mechanisms are quite wrong and would lose you marks. That is because curly arrows mean something quite specific. Explain what curly arrows represent and why the above drawings are wrong.

Answer

Curly arrows represent the movement of electrons. A double-headed arrow shows where a pair of electrons are going. In the right-hand drawing above the arrow would signify that the lone pair of electrons on oxygen is forming a bond to the proton—completely the opposite to what the student intended. In the left-hand drawing, the arrow would signify that a lone pair of electrons (which does not exist) on hydrogen is forming a bond to nothing at all!

5.5 Electrophilic additions of alkynes

Alkynes can undergo electrophilic additions with the same reagents as alkenes. Since there are two π bonds in alkynes, it is possible for the reaction to go once or twice depending on the amount of reagent added. Identify the product which you would expect if the following alkyne was treated with (a) 1 equivalent of Br_2, (b) 2 equivalents of Br_2.

$$H_3C - C \equiv C - CH_3$$

$$H_3C - C \equiv C - CH_3 \xrightarrow[\text{Br}_2]{\text{1 equiv.}}$$

$$H_3C - C \equiv C - CH_3 \xrightarrow[\text{Br}_2]{\text{2 equiv.}}$$

Answer

$$H_3C - C \equiv C - CH_3 \xrightarrow[\text{Br}_2]{\text{1 equiv.}}$$

(E) isomer

$$H_3C - C \equiv C - CH_3 \xrightarrow[\text{Br}_2]{\text{2 equiv.}}$$

With 1 equivalent of the reagent, an alkene is obtained. The *(E)* isomer is the observed product. With 2 equivalents of reagent, the alkene is formed first as an intermediate and then reacts further to give the tetrabromoalkane product.

Suggest the product which might be formed if the above alkyne was treated with 1 equivalent of HBr.

$$H_3C - C \equiv C - CH_3 \xrightarrow[\text{HBr}]{\text{1 equiv.}}$$

Answer

With 1 equivalent of reagent, electrophilic addition takes place once to give an alkene.

$$H_3C—C{\equiv}C—CH_3 \quad \xrightarrow[\text{HBr}]{\text{1 equiv.}} \quad \underset{\substack{H}}{\overset{\substack{H_3C}}{}}C{=}C\underset{\substack{CH_3}}{\overset{\substack{Br}}{}} \qquad \textit{(Z)} \text{ isomer only}$$

Suggest what might be formed if the above alkyne was treated with 2 equivalents of HBr.

$$H_3C—C{\equiv}C—CH_3 \qquad \longrightarrow$$

Answer

$$H_3C—C{\equiv}C—CH_3 \quad \xrightarrow[\text{HBr}]{\text{1 equiv.}} \quad \underset{\substack{H}}{\overset{\substack{H_3C}}{}}C{=}C\underset{\substack{CH_3}}{\overset{\substack{Br}}{}} \quad \xrightarrow[\text{HBr}]{\text{1 equiv.}} \quad H_3C—\underset{\substack{H}}{\overset{\substack{H}}{C}}—\underset{\substack{Br}}{\overset{\substack{Br}}{C}}—CH_3$$

Observed

Note that the final product has both bromines on the same carbon. The alternative product below is not formed.

$$H_3C—\underset{\substack{Br}}{\overset{\substack{H}}{C}}—\underset{\substack{H}}{\overset{\substack{Br}}{C}}—CH_3$$

Give the products which would be obtained if the unsymmetrical alkyne below was treated with (a) an excess of Br_2 and (b) an excess of HBr.

(a)

$$CH_3—C{\equiv}C—H \quad \xrightarrow{\text{Br}_2}$$

(b)

$$CH_3—C{\equiv}C—H \quad \xrightarrow{\text{HBr}}$$

Answer

(a)

$CH_3-C\equiv C-H$ $\xrightarrow{Br_2}$ [structure: CH_3 and Br on one carbon, Br and H on the other, C=C double bond]

Intermediate

$\xrightarrow{Br_2}$ [structure: $CH_3-C-C-H$ with Br, Br on first carbon and Br, Br on second carbon]

(b)

$CH_3-C\equiv C-H$ $\xrightarrow{HBr}$ [structure: CH_3 and H on one carbon, Br and H on the other, C=C double bond]

Intermediate

$\xrightarrow{HBr}$ [structure: $CH_3-C-C-H$ with Br, H on first carbon and H, H on second carbon]

Note that the reaction of the unsymmetrical alkyne with HBr gives the product shown and none of the alternative product below.

[structure: $CH_3-C-C-H$ with H, H on first carbon and Br, Br on second carbon]

This is another example of Markovnikov's rule whereby the hydrogens end up on the carbon which has the most hydrogens already present. We have already explained that this is due to the greater stability of one possible carbocation intermediate over the other.

The above addition reactions are similar to the addition reactions of alkenes. However, the reaction goes much more slowly for an alkyne than for an alkene which means that alkynes are less reactive than alkenes to electrophilic addition. This might seem strange since alkynes have four π electrons and you would think that they would be more electron-rich and more nucleophilic than an alkene with two π electrons. However, electrophilic addition to an alkyne involves the formation of a vinylic carbocation. This carbocation is much less stable than the carbocation intermediate resulting from electrophilic addition to an alkene. This is mostly due to the fact that less hyperconjugation is possible in the vinylic carbocation.

[structure: $(CH_3)(H_3C)C=C(H)(H)$] $\xrightarrow{HBr}$ [structure: carbocation with H_3C, H_3C on positively charged carbon, and H, H on other carbon]

Carbocation intermediate

$\xrightarrow{HBr}$ [structure: $H_3C-C-C-H$ with H_3C, :Br on first carbon, H, H on second carbon]

$H_3C-C\equiv C-H$ $\xrightarrow{HBr}$ [structure: $H_3C-C=C$ vinylic carbocation with positive charge]

Vinylic carbocation: unstable

$\xrightarrow{HBr}$ [structure: $(H_3C)(:Br)C=C(H)(H)$]

One consequence of this reduced reactivity is the reluctance of alkynes to react with acidified water. In order to get the reaction to take place, mercuric sulphate has to be added. The mercuric ion forms a complex with the triple bond and activates it for addition. Give the product you would expect if the following alkyne was treated with acidified water in the presence of mercuric sulphate.

$$CH_3 - C \equiv C - H$$

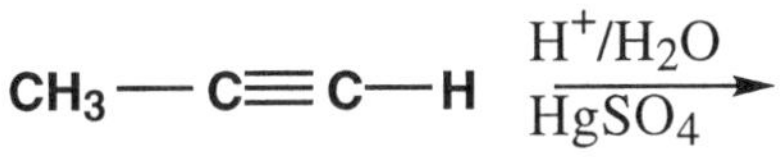

$$CH_3 - C \equiv C - H \xrightarrow[\text{HgSO}_4]{\text{H}^+/\text{H}_2\text{O}}$$

Answer

Taking into account Markovnikov's rule, you might have suggested the following reaction scheme:

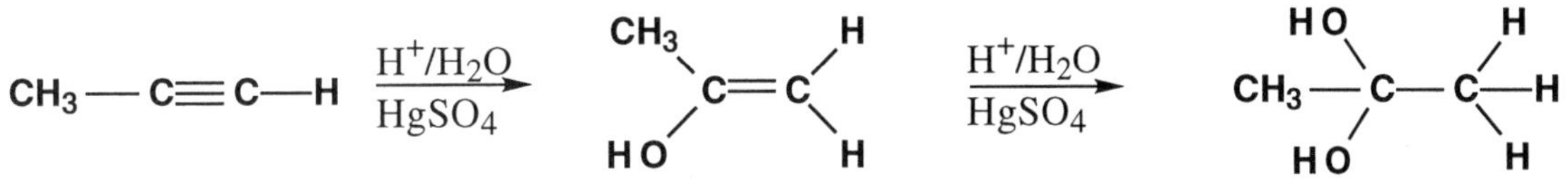

Enol intermediate

In fact this is not what happens. The intermediate gets 'diverted'.

The actual product obtained from this reaction is a ketone product which is formed by acid-catalysed rearrangement of the enol intermediate. The ketone and enol are in equilibrium with each other, with the equilibrium favouring the ketone.

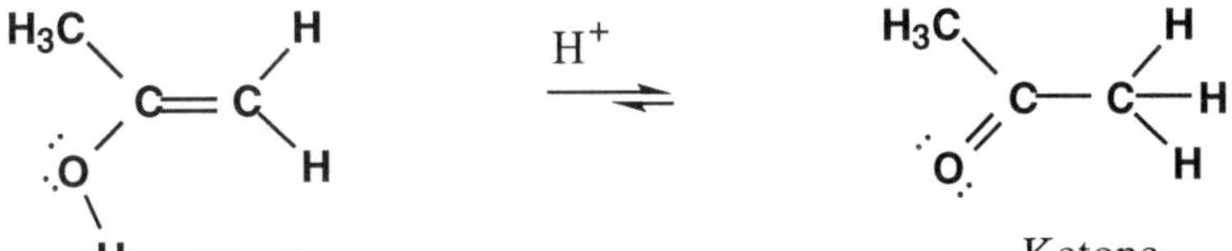

Enol   Ketone

The mechanism is as follows.

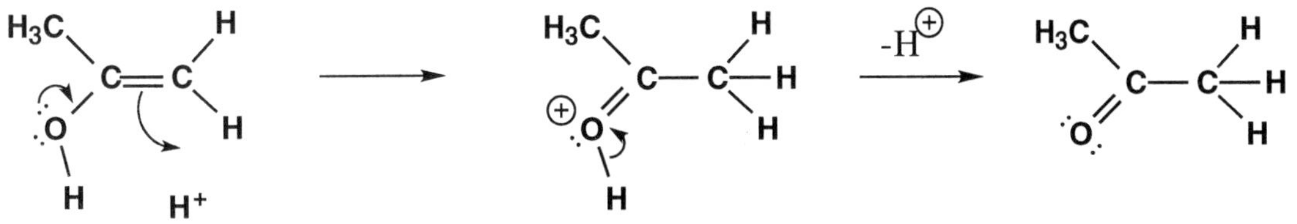

Explain this mechanism by describing which electrons are used in the making and breaking of bonds.

Answer

In the first step of the mechanism, one of the oxygen's lone pair of electrons is used to form a π bond between carbon and oxygen and makes a double bond. The oxygen gains a positive charge since it has effectively lost an electron. At the same time as the π bond is being formed, one of the bonds already present to carbon must be broken to avoid having five bonds to a carbon. The weakest bond will break, that is the π bond between the two carbons. Both electrons in this bond are used to form a bond to a proton.

In the second step, the two electrons forming the OH bond move onto the oxygen, thus breaking the OH bond. The hydrogen leaves the molecule as a proton. The two electrons form the second lone pair on oxygen and consequently neutralise the positive charge.

Summary

- Alkenes and alkynes undergo reactions known as **electrophilic additions**.

- Reagents which can be used for electrophilic additions are HBr, HCl, HI, Br_2, I_2, Cl_2 and H_2O.

- The reaction is called an electrophilic addition since the first step in the mechanism involves the addition of an electrophile to the alkene or the alkyne.

- The alkene π bond is used for the new bond to the electrophile since it is the weakest bond and the easiest to break.

- The mechanism for electrophilic addition has two steps involving a carbocation.

- Carbocations are stabilised by two effects. Alkyl substituents can donate electrons inductively. Neighbouring C–H bonds can stabilise the vacant 2p orbital by hyperconjugation.

- Markovnikov's rule states that a reagent such as HBr adds to an alkene such that the H attaches to the carbon with fewer alkyl substituents and the X attaches to the carbon with more alkyl substituents.

- Alkynes have two π bonds and electrophilic addition can take place twice if enough reagent is present.

- Alkynes are less reactive toward electrophilic addition than are corresponding alkenes.

- Reaction of alkynes with H_2O/H^+ requires $HgSO_4$ to be present. The product obtained is a ketone due to rearrangement of the enol obtained.

SECTION 6

Electrophilic Substitutions 1

Electrophilic Substitutions 1

6.1 Bromination

As we have seen, alkenes and alkynes can undergo electrophilic addition reactions. Aromatic rings, on the other hand, undergo electrophilic substitutions. One example of this reaction is the following bromination (at this stage we shall ignore how the bromine cation is formed).

Why is this reaction known as electrophilic substitution?

Answer

The reaction involves Br^+ which is an electrophile. The Br^+ replaces a proton on the aromatic ring with the aromatic ring remaining intact. The reaction is an electrophilic substitution since one electrophile (Br^+) substitutes another (H^+).

We now need to work out the mechanism by which this reaction takes place. Identify the nucleophile and the electrophile when an aromatic ring reacts with Br^+.

Answer

The Br^+ ion is positively charged, electron-deficient and is therefore an electrophile. The aromatic ring has six π electrons and is therefore electron-rich and a nucleophile.

If the aromatic ring is to form a bond to an electrophile, two electrons are required. Where will they come from?

Answer

Any two of the aromatic ring's six π electrons could be used to form the new bond. When writing a mechanism, just choose one of the three 'double' bonds in the aromatic structure.

Show how an aromatic ring could form a bond to an electrophile such as Br^+ and draw the structure which would be obtained.

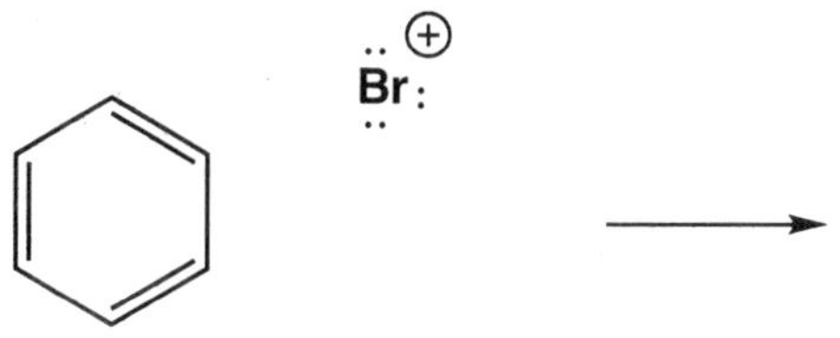

Answer

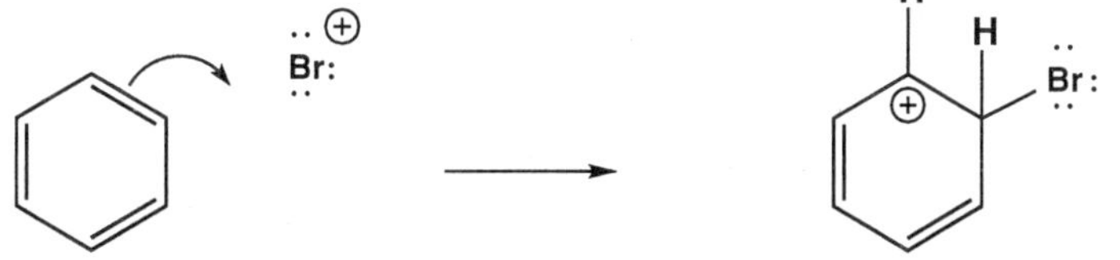

(The hydrogens shown here were on the original aromatic ring.)

Two of the π electrons in the aromatic ring are used to form a bond to the bromine ion. In order to do that, the aromatic ring loses one of its double bonds and a carbon atom gains a positive charge. This is exactly the same mechanism which we saw in the first stage of electrophilic additions to alkenes.

The positively charged intermediate here is equivalent to the positively charged intermediate in electrophilic additions.

Intermediate in the
electrophilic addition
of alkenes.

Intermediate in the
electrophilic substitution
of aromatics.

The mechanism involving the initial addition of the electrophile to an alkene or an aromatic ring is the same. However, the mechanism then changes. Whereas the intermediate from an alkene reacts with a nucleophile to give an addition product, the intermediate from the aromatic ring loses a proton and the aromatic ring is regenerated. Show a mechanism by which this can take place. Remember you have to draw an arrow which will reform the aromatic ring, neutralise the positive charge in the ring and produce a proton.

Answer

The two electrons in the carbon–hydrogen bond move to form a π bond in the ring and thus regenerate the aromatic ring and neutralise the positive charge. Since the electrons have been taken away from the hydrogen, a proton is produced which leaves the molecule.

The full mechanism is as follows.

Cationic
intermediate

The intermediate can be stabilised by 'spreading' the positive charge to different parts of the ring by resonance. Suggest how this takes place by considering the remaining π electrons in the ring.

Answer

Two of the remaining π electrons in the ring can move to neutralise the positive charge at the 'top' carbon. In doing so, a positive charge is generated at a different carbon. This can happen again such that the charge is spread between three carbons. This spreading of charge stabilises the positive charge.

These structures are known as resonance structures. The double headed arrow between the resonance structures indicates an easy interconversion between all three resonance structures.

We have shown how aromatic rings undergo electrophilic substitution rather than electrophilic addition, but we have not yet explained why. The reason lies in the extra stabilisation inherent in an aromatic system. The six π electrons in an aromatic ring are delocalised around the ring and this produces a stabilisation which would not be observed for three separate double bonds. With aromatic rings, electrophilic substitution is favoured over electrophilic addition because the stable aromatic ring is regenerated. This would not be the case in electrophilic addition.

This extra stabilisation of an aromatic ring also means that aromatic compounds are less reactive than alkenes to electrophiles. For example, an alkene will react with Br_2 whereas an aromatic ring will not. This poses a problem when it comes to the bromination of the aromatic ring above. Up until now we have used a bromine cation without explaining where it has come from. Bromine cations are too unstable to be formed in isolation and have to be generated from Br_2 during the mechanism. We saw this in the reaction of Br_2 with alkenes earlier on.

In the reaction of Br_2 with alkenes, the electron-rich, nucleophilic double bond polarises the Br-Br bond, then forms a bond to the electrophilic bromine. At the same time, the Br-Br bond is broken. In this way, reaction takes place without the need for a free bromine cation.

However, the aromatic ring is not as strong a nucleophile as an alkene and is unable to carry out this first step on Br_2 alone. In order to get this reaction to occur we either have to activate the aromatic ring (i.e. make it a better nucleophile) or activate the Br_2 (i.e. make it a better electrophile). In a later section, you will see how electron-donating substituents on an aromatic ring increase the nucleophilicity of the aromatic ring. At this stage, we shall see how we can activate the Br_2 molecule to make it a better electrophile.

The method used is to add a Lewis acid to the reaction medium. Examples of Lewis acids include $FeCl_3$, $FeBr_3$, and $AlCl_3$. These Lewis acids all contain a central atom (iron or aluminium) which is strongly electrophilic, so much so that it can accept a lone pair of electrons from a weakly nucleophilic atom such as a halogen. The Lewis acid used to catalyse bromination of an aromatic ring is $FeBr_3$. One of the bromine atoms in Br_2 will use a lone pair to bond to $FeBr_3$. Draw a mechanism to show this and draw the structure which is obtained.

:Br—Br: FeBr3   ⟶

Answer

:Br—Br:⤴   FeBr3   ⟶   ⊕Br—Br⊖—FeBr3

Both electrons for the new bond come from bromine. Therefore, the bromine which bonds to iron effectively loses an electron and becomes positively charged.

The iron effectively gains an electron and becomes negatively charged.

By bonding to a Lewis acid, bromine is activated towards a nucleophile and will react as follows.

Nu ⟶   :Br⊕—Br⊖—FeBr3   ⟶   Nu—Br: :Br⊖—FeBr3

When bromine bonds to a Lewis acid, one of the bromine atoms gains a positive charge. This makes bromine highly reactive since a positive charge on a halogen is not stable. The aromatic ring can now react with the activated bromine. Draw a mechanism to show how the aromatic ring will react with the bromine/Lewis acid complex.

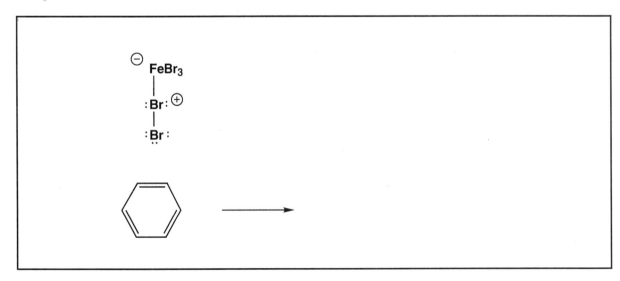

Answer

Two of the π electrons in the aromatic ring are used to form a bond to bromine. At the same time as this bond is formed, the bromine–bromine bond is broken with both electrons being used to form a new lone pair on the positively charged bromine. As a result, the bromine loses its positive charge. The second stage of the mechanism is as before.

The aromatic ring can also be chlorinated using Cl_2 in the presence of a Lewis acid ($FeCl_3$).

Chlorination

$$\frac{Cl_2}{FeCl_3}$$

6.2 Friedel–Crafts alkylation and acylation

Alkylation and acylation is also possible using $AlCl_3$ as the Lewis acid.

Alkylation

$$\frac{R–Cl}{AlCl_3}$$

R = alkyl group
(e.g. CH_3, CH_2CH_3)

Acylation

$$\frac{}{AlCl_3}$$

R=alkyl group
(e.g. CH_3, CH_2CH_3)

A Lewis acid is necessary for this reaction and promotes the formation of the carbocations required for these reactions. It does so by accepting a lone pair of electrons from the chlorine atom of the alkyl halide or acyl halide.

Draw the mechanism for the reaction of $AlCl_3$ with the following alkyl halide.

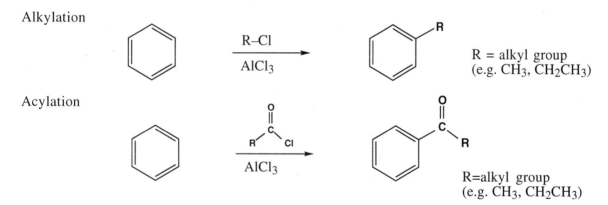

Answer

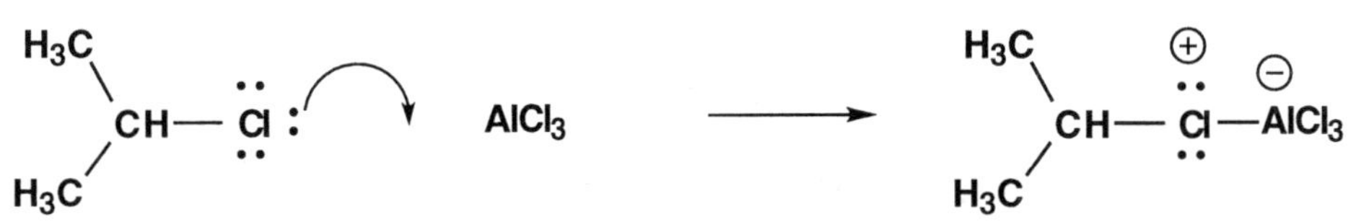

$$H_3C \diagdown CH - \overset{..}{\underset{..}{Cl}} : \quad \overset{\curvearrowright}{} \quad AlCl_3 \quad \longrightarrow \quad H_3C \diagdown CH - \overset{\oplus}{\underset{..}{Cl}} \overset{\ominus}{-} AlCl_3$$

This product is very unstable due to the positively charged chlorine and so the structure fragments to give a carbocation and $AlCl_4^-$. Draw arrows in the following diagram to show the mechanism for this.

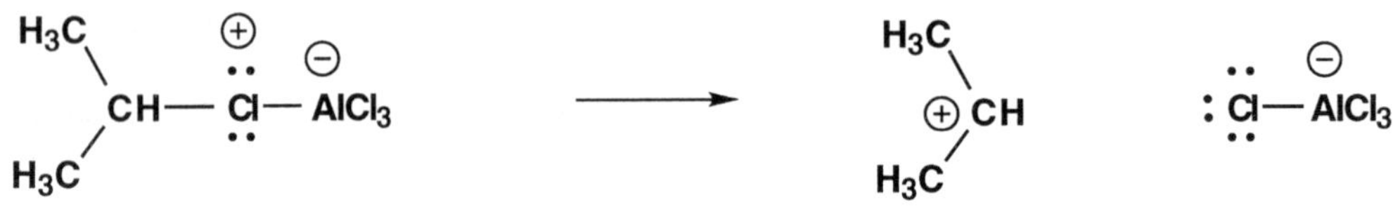

$$H_3C \diagdown CH - \overset{\oplus}{\underset{..}{Cl}} \overset{\ominus}{-} AlCl_3 \quad \longrightarrow \quad H_3C \diagdown \overset{\oplus}{CH} \quad \overset{..}{\underset{..}{Cl}} \overset{\ominus}{-} AlCl_3$$

Answer

$$H_3C \diagdown CH - \overset{\oplus}{\underset{..}{Cl}} \overset{\ominus}{-} AlCl_3 \quad \longrightarrow \quad H_3C \diagdown \overset{\oplus}{CH} \quad : \overset{..}{\underset{..}{Cl}} \overset{\ominus}{-} AlCl_3$$

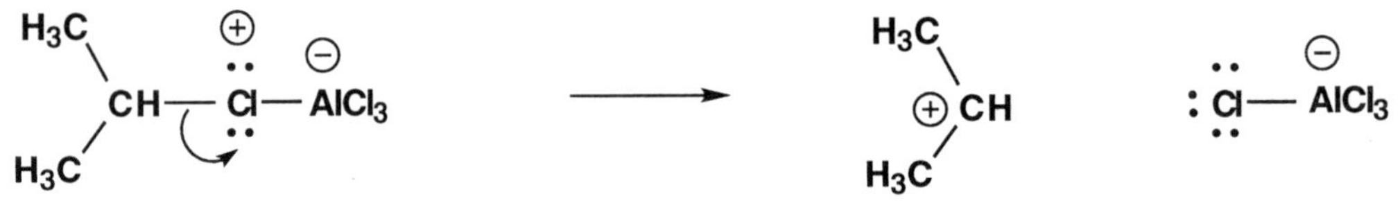

Once the carbocation is formed, it can react with the aromatic ring.

The acyl carbocation is generated in the same way. Show the mechanism by which the carbocation of propanoyl chloride is formed.

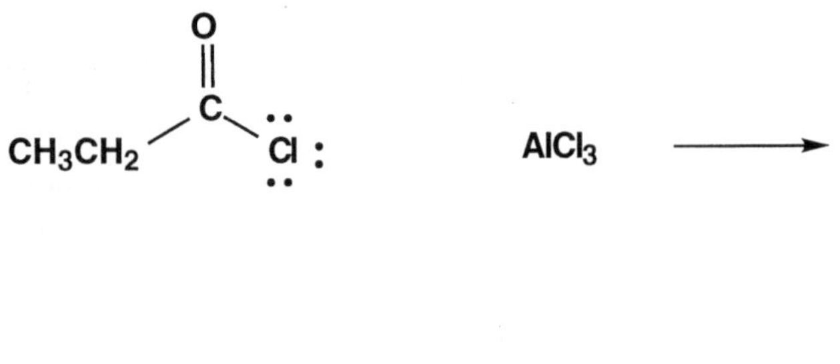

$$CH_3CH_2 \diagdown \underset{\parallel}{\overset{O}{C}} \diagdown \overset{..}{\underset{..}{Cl}} : \quad AlCl_3 \quad \longrightarrow$$

Answer

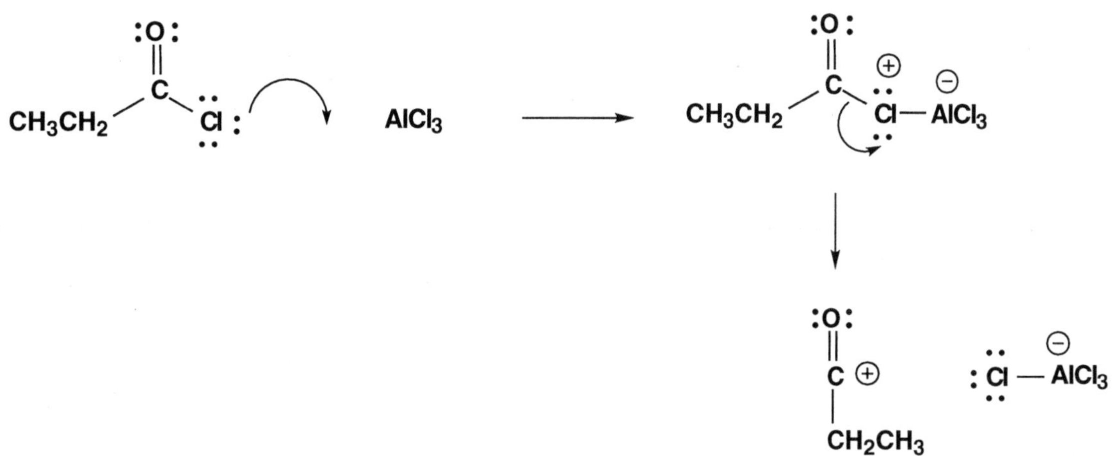

This ion is known as the acylium ion. As drawn above, it has the positive charge localised on the carbon atom. This would be quite unstable, and it would be difficult to form the ion. However, the charge can be stabilised by delocalising the charge, such that it is spread between carbon and oxygen. Draw a mechanism to demonstrate how this can happen, and draw the resulting resonance structure.

Answer

One of the lone pairs on oxygen can form a second π bond to carbon. This results in oxygen gaining the positive charge and carbon losing the charge. The actual structure of the acylium ion will be a resonance hybrid of these two extremes. Since the charge is delocalised, it is more stable, and the ion is easier to form.

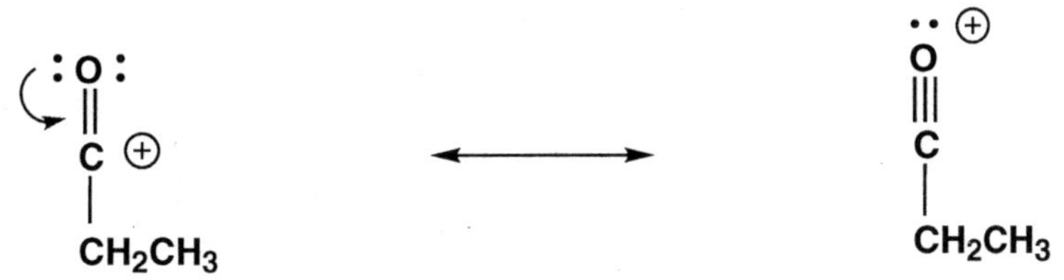

Give the product you would expect if benzene is treated with 2-chloropropane and $AlCl_3$. Show also the mechanism of the reaction.

Show the product.

Draw the mechanism for the generation of the carbocation.

Draw the mechanism for the reaction of the carbocation with benzene.

Answer

The product of the reaction is 2-phenylpropane.

2-Phenylpropane

The mechanisms are as follows.

Generation of the carbocation.

Explanation:
The chlorine in the alkyl halide is a weak nucleophile, but is capable of forming a bond to the electrophilic centre of $AlCl_3$ (the aluminium atom). Chlorine provides both electrons for the new bond and gains a positive charge. This is not very stable and so the C–Cl bond breaks in order to neutralise the charge. Both electrons move onto chlorine resulting in the formation of the carbocation.

The reaction of the carbocation with benzene.

Explanation:
Stage 1
The carbocation is obviously an electrophile.
The aromatic ring acts as a nucleophile supplying two of its π electrons to form a bond to the electrophilic carbon on the carbocation. As a result, the aromatic ring gains a positive charge.

Stage 2
The ring is now positively charged and can be viewed as an electrophile. Potentially, it could react with an 'external' nucleophile in the same way as an alkene. However, this would mean loss of aromaticity which in turn means a loss of stability. Instead, a proton is lost. Both electrons which make up the carbon–hydrogen bond are used to restore a total of six π electrons in the ring. The hydrogen is lost as a proton.

Explain with mechanisms how the reaction intermediate in the problem above can be stabilised, thus making the reaction more likely.

Answer

The reaction intermediate is positively charged and is therefore reactive. If the positive charge can be 'spread' or delocalised around the ring, this will help to stabilise the intermediate. As a result, it is more likely to be formed and consequently, the overall reaction is more likely to take place. The mechanism by which the positive charge can be delocalised involves a movement of the remaining π electrons in the ring as shown below.

A common mistake in drawing mechanisms is to be careless in drawing the curly arrows. For example, the mechanism below was given as an answer to the above question in a class test. The student was not given full marks. Why not?

Answer

There are two basic errors in the above answer. First of all, the student did not use double-headed arrows between the resonance structures. The double-headed arrows are important in showing that there is a rapid interconversion of these structures in both directions. Secondly, the two curly arrows are badly drawn. They both start from the right place (the location of two π electrons) but they point to the wrong place. Both arrows point to the carbon with the positive charge. This would mean that both electrons are moving onto that carbon atom.

However, that is not what happens. The two electrons are being used to form a π bond between two carbons and so the arrow must point to where the middle of the new bond will be formed. Curly arrows must always be drawn to where a new bond is being formed. The only time you draw an arrow to an atom is when both electrons are forming a lone pair on that atom.

There are limitations to the Friedel–Crafts alkylation. For example, the reaction of 1-chlorobutane with benzene gives a mixture of products.

34%   66%

This is due to the fact that the primary carbocation which is generated can rearrange to the secondary carbocation by the following mechanism.

In this mechanism, the hydrogen is 'shifted' across to the neighbouring carbon atom along with the two electrons in the C–H bond. This is known as a hydride shift. Let us now see why this happens.

Compare the two carbocations above. Would you expect them to be equally stable? (Remember hyperconjugation and the inductive effect of alkyl groups.)

Answer

The secondary carbocation is more stable than the primary carbocation since there are more electron-donating alkyl groups attached to the positive centre. The inductive effect lowers the positive charge and helps to stabilise it.

Stabilised by one   Stabilised by two
alkyl group   alkyl groups

Hyperconjugation between the empty 2p orbital of the carbocation and neighbouring C–H bonds is more significant in the secondary carbocation. There are five available C–H σ bonds in the secondary carbocation as opposed to two in the primary carbocation.

Rearrangement therefore leads to a more stable carbocation. Such rearrangements will occur whenever it is possible to obtain a more stable secondary or tertiary carbocation from a primary carbocation.

Show how the following carbocation could rearrange to give a more stable ion.

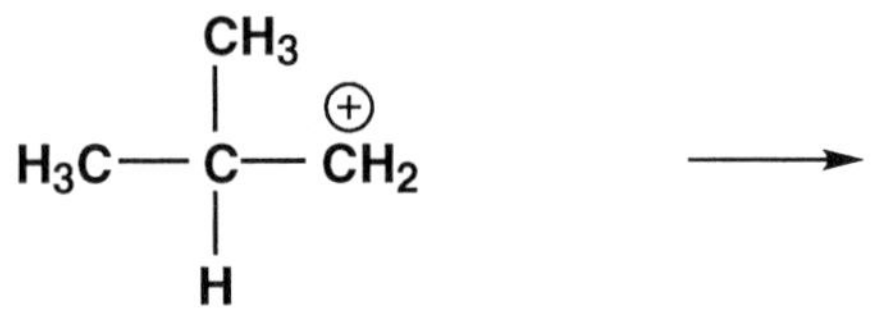

Answer

This rearrangement reaction limits the type of alkylations which you can do. For example, consider the following reactions. Identify those which are likely to work well and those which are likely to give side products due to carbocation rearrangements. Show the structures of the alternative products where appropriate.

(a)

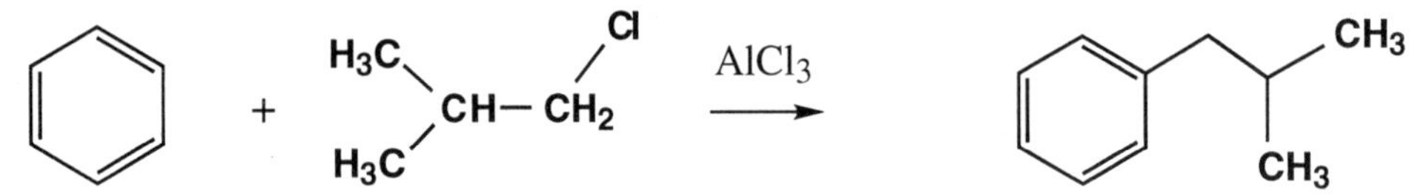

(b)

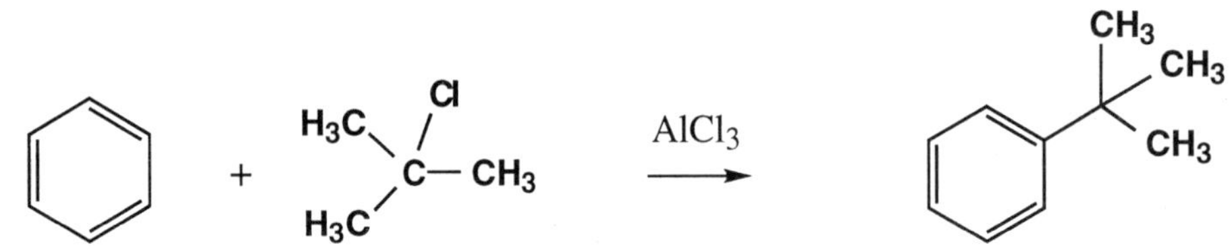

(c)

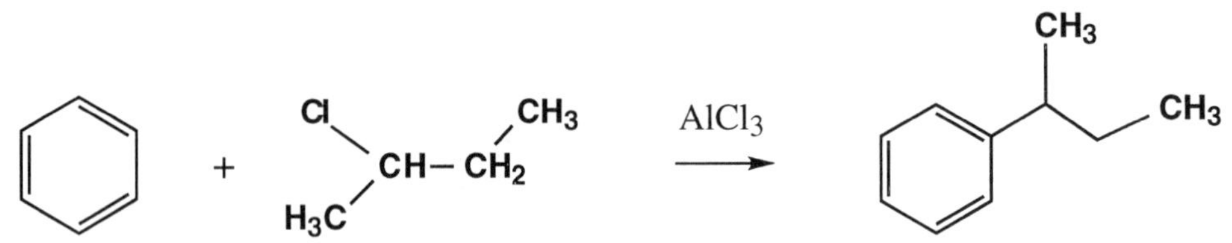

(d)

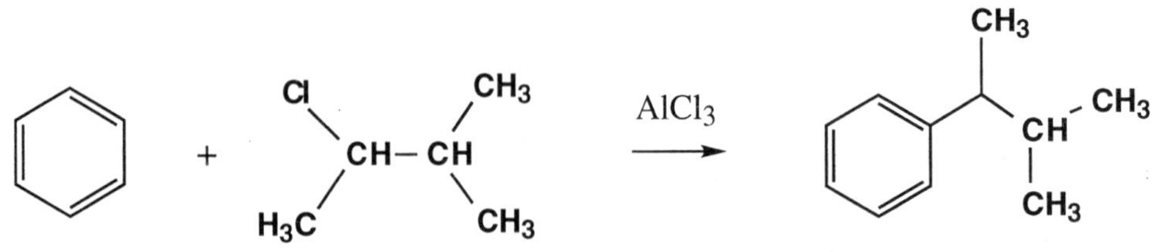

Answer

Reactions (b) and (c) are likely to be successful since the carbocations involved cannot rearrange to a more stable carbocation. Reactions (a) and (d) are likely to be unsuccessful since rearrangement to a more stable carbocation is possible. The alternative products in (a) and (d) are the following.

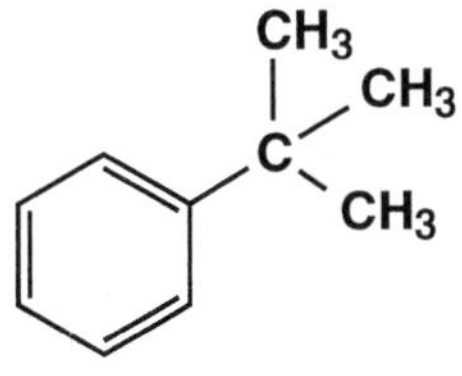

Alternative product
(reaction a)

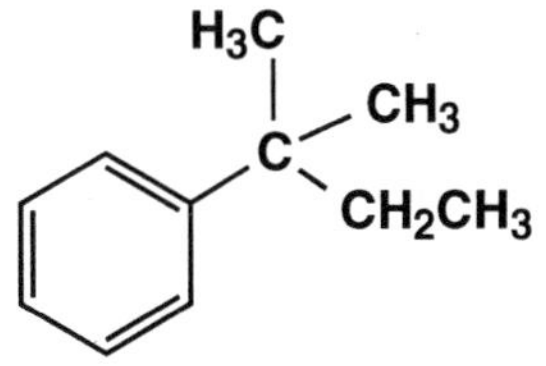

Alternative product
(reaction d)

Show the mechanism by which the carbocations originally formed in reactions (a) and (d) can rearrange to give the carbocations leading to these alternative products.

(a)

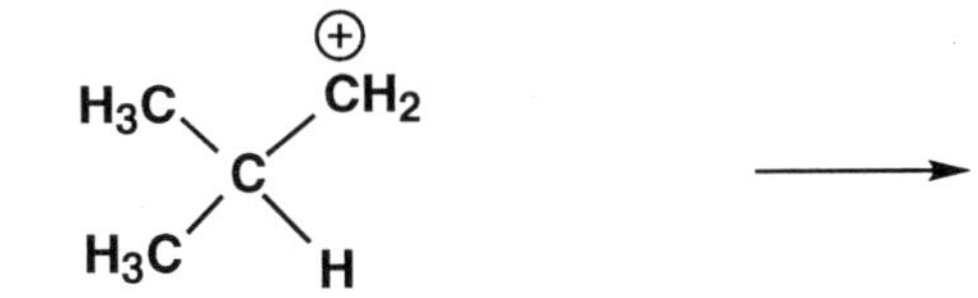

(b)

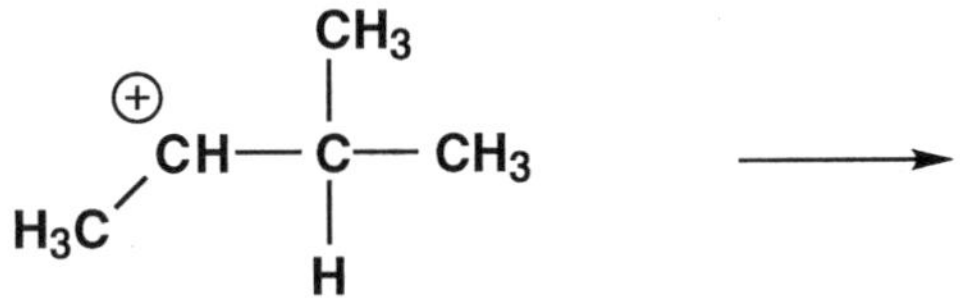

Answer

(a)

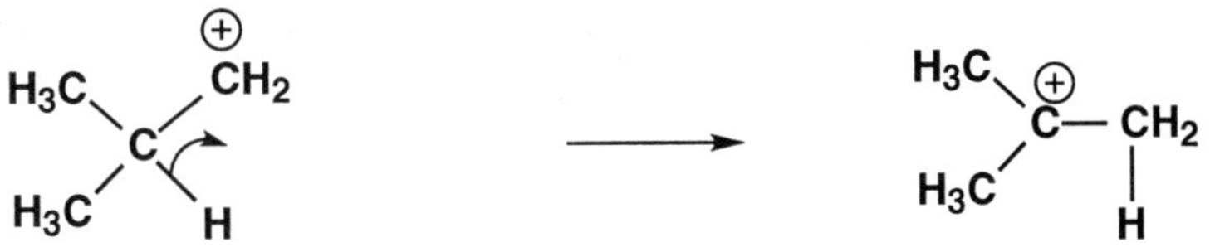

(b)

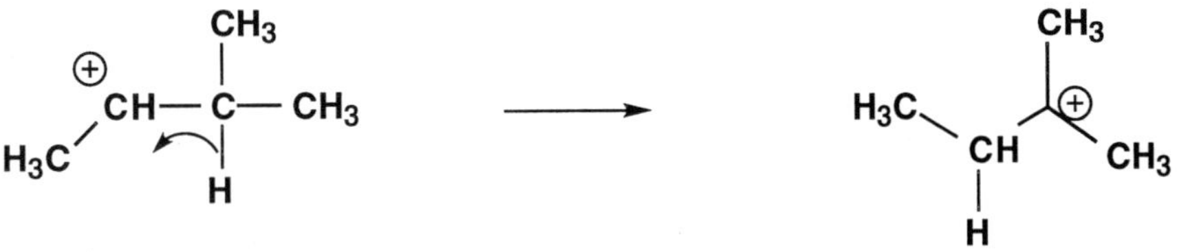

How then is it possible to prepare a compound such as the following?

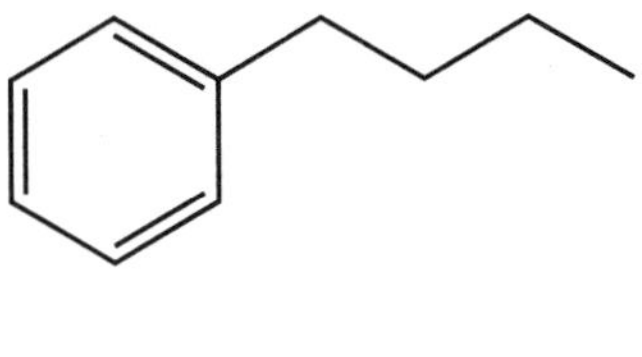

Friedel–Crafts alkylation using 1-chlorobutane is not very efficient due to rearrangement of the carbocation intermediate. The answer is to carry out a Friedel–Crafts acylation to attach the four-carbon skeleton. This is because the acyl carbocation does not rearrange. Once the acyl group is attached it is possible to reduce the keto functional group using zinc/HCl or with NH_2NH_2/NaOH.

Example

$$\text{benzene} \xrightarrow[\text{AlCl}_3]{\text{CH}_3\text{CH}_2\text{CH}_2\text{COCl}} \text{phenyl butyl ketone} \xrightarrow{\text{Zn/HCl}} \text{C}_6\text{H}_5\text{–CH}_2\text{CH}_2\text{CH}_2\text{CH}_3}$$

Show how you would synthesise the following aromatic structures.

isobutylbenzene   *tert*-butylbenzene

---

benzene $\longrightarrow$

benzene $\longrightarrow$

Answer

$$\text{benzene} \xrightarrow[\text{AlCl}_3]{(\text{CH}_3)_2\text{CHCOCl}} \text{(keto intermediate)} \xrightarrow{\text{Zn/HCl}} \text{isobutylbenzene}$$

$$\text{benzene} \xrightarrow[\text{AlCl}_3]{(\text{CH}_3)_3\text{CCl}} \textit{tert}\text{-butylbenzene}$$

So far we have looked at Friedel–Crafts alkylation using an alkyl halide with a Lewis acid such as AlCl₃. However, it is also possible to carry out Friedel–Crafts alkylations using alcohols or alkenes in the presence of an acid. Treatment of these functional groups with acid leads to carbocations which can then react with an aromatic ring by the same electrophilic substitution mechanism we have studied.

You have already seen in Section 5 how alkenes react with an acidic proton to give a carbocation. The reaction of alcohols with acid leads to the elimination of water to give an alkene which can then be converted to a carbocation in the same way.

Show how the following alkene would react with a proton to give a carbocation and show what product would be formed if that carbocation was treated with benzene.

The conversion of alcohols to alkenes under acid conditions will be covered in Section 12 of this text.

6.3 Sulphonation and nitration

There are two other common electrophilic substitutions which can be carried out on aromatic rings. These involve strong electrophiles and do not need the presence of a Lewis acid.

Sulphonation

Nitration

In sulphonation, the electrophile is sulphur trioxide (SO_3). Draw the structures of sulphuric acid and sulphur trioxide.

Answer

Sulphuric acid   Sulphur trioxide

The mechanism by which sulphur trioxide is generated from sulphuric acid involves acid catalysis via a protonated intermediate (I).

Put in the arrows for the mechanism.

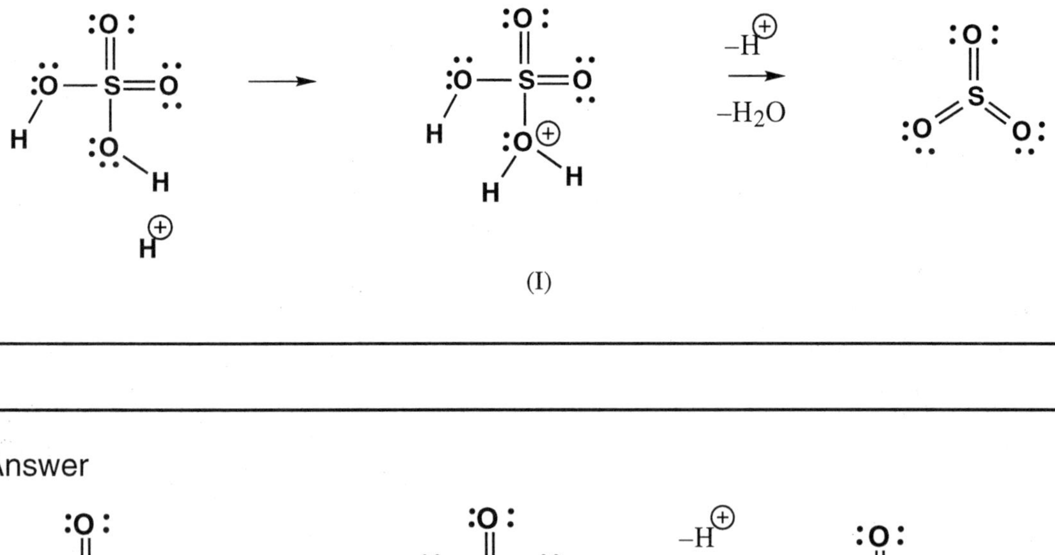

(I)

Answer

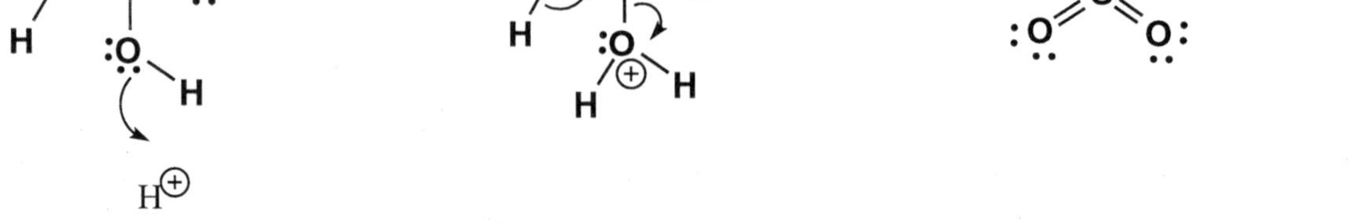

Acid catalysis involves the protonation of an **OH** group on sulphuric acid such that it eventually leaves as water. Why should protonation help this process?

Answer

By protonating the OH group, the oxygen gains a positive charge and this makes it a much better leaving group.

Sulphur trioxide has no positive charge. Why should it be an electrophile?

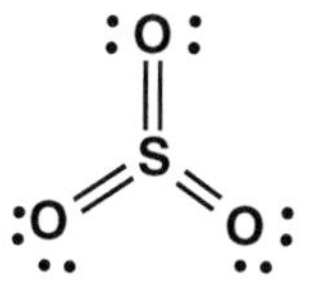

Answer

The sulphur atom is bonded to three electronegative oxygen atoms which are 'pulling' electrons from the sulphur and making it electron-deficient (i.e. electrophilic).

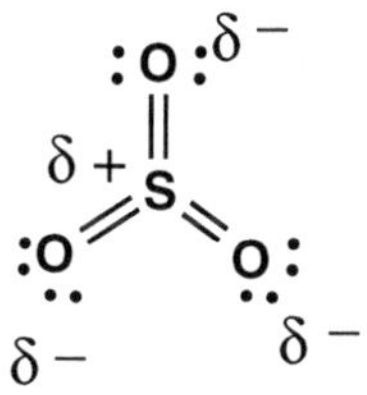

Now that we have identified the electrophilic centre of sulphur trioxide, draw the mechanism for the sulphonation of benzene. (Note that sulphur is not allowed more than six bonds).

Answer

The mechanism as far as the aromatic ring is concerned is exactly the same as before. Two π electrons are used to form a bond to the electrophilic centre of the electrophile (i.e. sulphur). A positively charged intermediate is formed and a proton is lost to regenerate the aromatic ring.

As far as the electrophile is concerned (i.e. SO_3), the sulphur atom is only allowed six bonds, so when the new bond is formed to sulphur, one of the bonds already present must break. The easiest bond to break is one of the π bonds between sulphur and oxygen. Both electrons will move to the more electronegative oxygen to form a third lone pair and a negative charge on that oxygen. This will finally be neutralised when the third lone pair is used to form a bond to a proton.

In nitration, nitric acid and sulphuric acid are used. The sulphuric acid serves as an acid catalyst to generate a nitronium ion ($^{\oplus}NO_2$). Draw the structures for nitric acid and the nitronium ion.

Answer

The nitronium ion is generated from nitric acid by a very similar mechanism to that used in the generation of sulphur trioxide from sulphuric acid. Try and write this mechanism, remembering that a proton helps to catalyse the reaction.

Draw the mechanism for the nitration of benzene. (Note that nitrogen is not allowed more than four bonds.)

Answer

The mechanism is very similar to sulphonation. As the aromatic ring forms a bond to the electrophilic nitrogen atom, one of the four bonds already present to nitrogen must break. The easiest bond to break is one of the π bonds, with both electrons moving onto the oxygen atom. Unlike sulphonation, this oxygen keeps its negative charge and does not pick up a proton. This is because it is acting as a counterion to the neighbouring positive charge on nitrogen.

6.4 Aromatic substituents by functional group transformation

Some substituents cannot be introduced directly onto an aromatic ring by electrophilic substitution. These include the following groups:

NH_2, NHR, NR_2, $NHCOCH_3$, CO_2H, CN, OH

These groups are obtained by transforming a functional group which **can** be applied directly by electrophilic substitution. The most important transformations are as follows.

Nitro, alkyl and acyl groups can readily be added by electrophilic substitution and can be converted to amino, carboxylic acid and alkyl groups respectively. It may not seem obvious why the last reaction is useful since alkyl groups can be added directly by the Friedel–Crafts alkylation. However, as we saw earlier, there are some alkyl groups which rearrange on Friedel–Crafts alkylation and it is better to carry out a Friedel–Crafts acylation followed by a reduction.

The amino group and the carboxylic acid group can be converted to a large range of other functional groups. A particularly useful reaction is the treatment of the amino group with sodium nitrite ($NaNO_2$). This gives an intermediate known as a diazonium salt which can be converted to a wide variety of other substituents.

$$Ar\!-\!NH_2 \xrightarrow{\text{1 equiv. R–I}} Ar\!-\!NHR$$

$$\xrightarrow{\text{2 equiv. R–I}} Ar\!-\!NR_2$$

$$\xrightarrow[\text{acid chloride}]{\text{RCOCl}} Ar\!-\!NHCOR \xrightarrow{\text{LiAlH}_4} Ar\!-\!NHCH_2R$$

$$\xrightarrow[\text{(2) KI}]{\text{(1) NaNO}_2/\overset{\oplus}{\text{H}}} Ar\!-\!I$$

$$\xrightarrow[\text{(2) CuCN}]{\text{(1) NaNO}_2/\overset{\oplus}{\text{H}}} Ar\!-\!CN \xrightarrow{\overset{\oplus}{\text{H}}/\text{H}_2\text{O}} Ar\!-\!CO_2H$$

$$\xrightarrow[\text{(2) CuCl}]{\text{(1) NaNO}_2/\overset{\oplus}{\text{H}}} Ar\!-\!Cl$$

$$\xrightarrow[\text{(2) CuBr}]{\text{(1) NaNO}_2/\overset{\oplus}{\text{H}}} Ar\!-\!Br$$

$$\xrightarrow[\text{(2) HBF}_4/\text{heat}]{\text{(1) NaNO}_2/\overset{\oplus}{\text{H}}} Ar\!-\!F$$

$$\xrightarrow[\text{(2) H}_2\text{O heat}]{\text{(1) NaNO}_2/\overset{\oplus}{\text{H}}} Ar\!-\!OH \xrightarrow[\text{2. RI}]{\text{1. NaOH}} Ar\!-\!OR$$

$$\xrightarrow{\text{RCOCl}} Ar\!-\!OCOR$$

A carboxylic acid can also be converted to other functional groups.

$$Ar\!-\!CO_2H \xrightarrow{\text{SOCl}_2} Ar\!-\!COCl \longrightarrow \text{anhydrides, esters and amides}$$

$$\xrightarrow{\overset{\oplus}{\text{H}}/\text{ROH}} Ar\!-\!CO_2R$$

$$\xrightarrow{\text{LiAlH}_4} Ar\!-\!CH_2OH$$

Show how you would synthesise the following aromatic compound from benzene.

Answer

When answering a question like this, it is best to work backwards from the product and ask yourself what it could have been synthesised from.

For example, in this case we have an ester attached to an aromatic ring. An ester functional group cannot be attached directly by electrophilic substitution, so the synthesis must involve several steps.

Working backwards, how could an ester be formed? The usual way to make an ester is from an acid chloride which is synthesised in turn from a carboxylic acid. Alternatively, the ester can be made directly from the carboxylic acid by treating it with an alcohol and an acid catalyst.

Either way, benzoic acid is the most likely intermediate in the synthesis.

The question now is how does one get benzoic acid? Carboxylic acids cannot be added directly to aromatic rings either, so we have to look for a different functional group or substituent which can be added directly, then transformed to a carboxylic acid. From the reactions given, you will see that a methyl group can be oxidised to a carboxylic acid. Methyl groups can be added directly by Friedel–Crafts alkylation. A possible synthetic route would be the following.

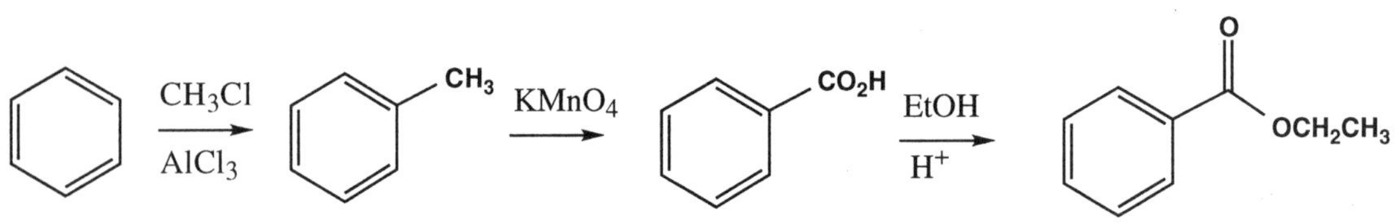

One problem with the above route would be the possibility of poly-methylation in the first step. This is likely since the product (toluene) will be more reactive than the starting material (benzene). One way round this problem would be to use an excess of benzene. Such a tactic is perfectly feasible for the first step in the reaction sequence, but would be extremely wasteful if the problem was encountered at a later stage.

There are other possible routes to the intermediate carboxylic acid but they are longer and the fewer steps in a synthetic sequence, the better.

Devise a synthetic route to the following structure.

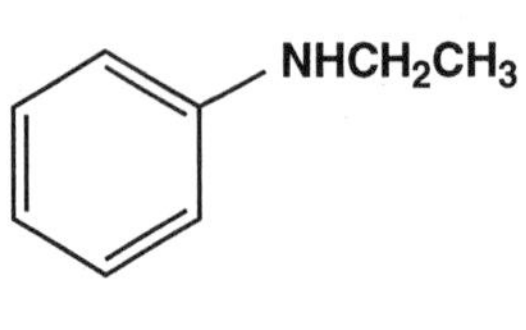

Answer

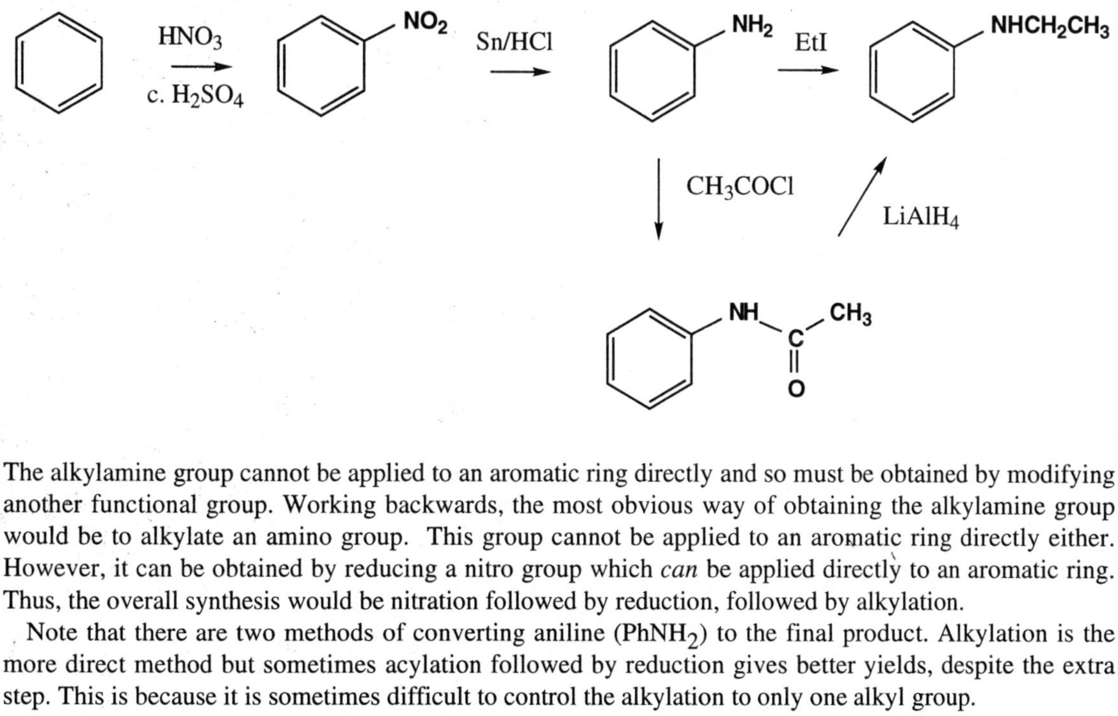

The alkylamine group cannot be applied to an aromatic ring directly and so must be obtained by modifying another functional group. Working backwards, the most obvious way of obtaining the alkylamine group would be to alkylate an amino group. This group cannot be applied to an aromatic ring directly either. However, it can be obtained by reducing a nitro group which *can* be applied directly to an aromatic ring. Thus, the overall synthesis would be nitration followed by reduction, followed by alkylation.

Note that there are two methods of converting aniline ($PhNH_2$) to the final product. Alkylation is the more direct method but sometimes acylation followed by reduction gives better yields, despite the extra step. This is because it is sometimes difficult to control the alkylation to only one alkyl group.

Devise a synthetic route to the following structure:

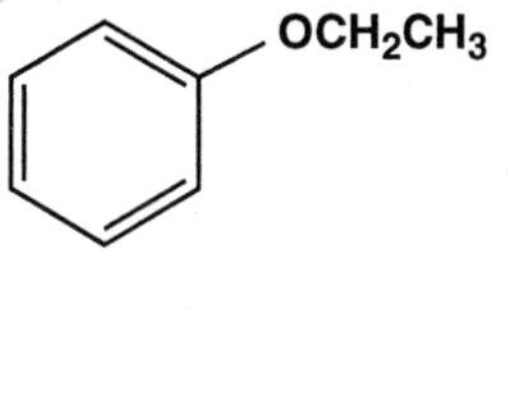

Answer

The ethyl ether cannot be applied directly to an aromatic ring, so we have to find a way of obtaining it from another functional group. Alkylation of a phenol group would give the desired ether.

We now have to work out how to get the phenol group since this cannot be applied directly to the ring either. Working backwards again, we can obtain the phenol from an amino group, which in turn can be obtained from a nitro group. This group can be applied directly to the ring and so the synthesis involves a nitration, reduction, conversion of the amino group to a diazonium salt, hydrolysis and finally an alkylation.

Summary

- Aromatic rings prefer electrophilic substitution over electrophilic addition since the former reaction regenerates the stabilised aromatic ring system.

- Aromatic rings are less reactive towards electrophiles than alkenes (unless they contain strongly activating substituents—see Section 7).

- Lewis acids are required to activate electrophiles for halogenation, alkylation and acylation.

- The most common electrophilic substitutions are halogenation, Friedel–Crafts alkylation, Friedel–Crafts acylation, nitration and sulphonation.

- A nitro group can be reduced to an amino group which can then be converted to a large range of other functional groups.

- An alkyl group can be oxidised to a carboxylic acid which can then be converted to other functional groups.

SECTION 7

Electrophilic Substitutions 2

Electrophilic Substitutions 2

In Section 6 we looked at the possible electrophilic substitutions for benzene. In all these cases, only one product was possible. When we start looking at the electrophilic substitution of aromatic compounds which contain one or more substituents, the picture becomes more complicated. A mixture of products is now possible depending on where the substitution takes place on the ring.

7.1 1,2-(ortho), 1,3-(meta) and 1,4-(para) substitution

Take, for example, toluene. How many different products might be possible if toluene is brominated? Draw their structures.

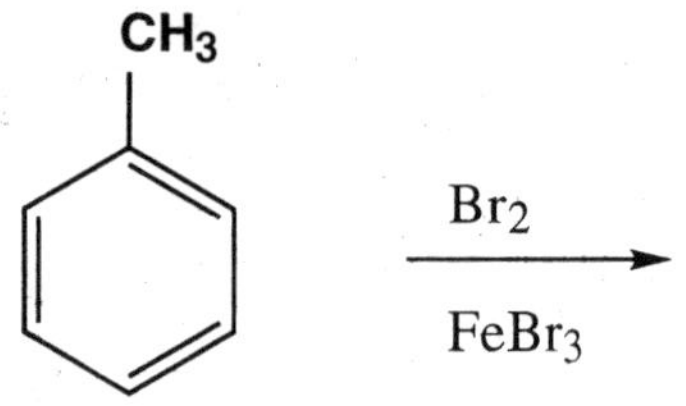

Toluene

Answer

Three different products are possible, depending on which part of the ring is substituted.

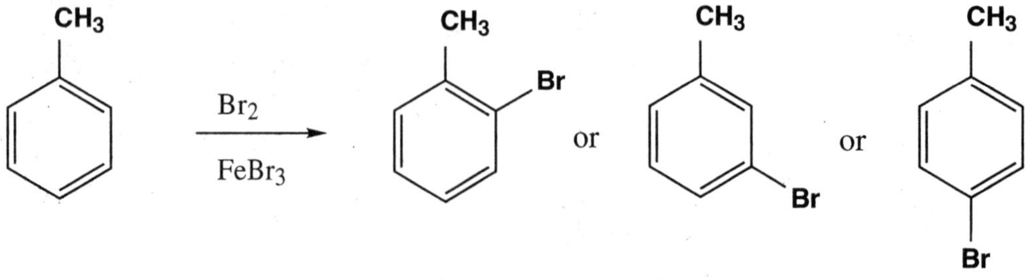

ortho   meta   para

These three different products have the same molecular formula and are therefore constitutional isomers. The aromatic ring is said to be disubstituted and the three possible isomers are described as being *ortho*, *meta* and *para*. This corresponds to 1,2-, 1,3-, and 1,4- substitution respectively.

Write the three mechanisms leading to these three isomers. (You should already know the basic mechanism from Section 6. All you need to do is repeat the mechanism such that the bromine is attached at the relevant positions. To simplify matters, assume the presence of the Br^+ ion.)

ortho substitution

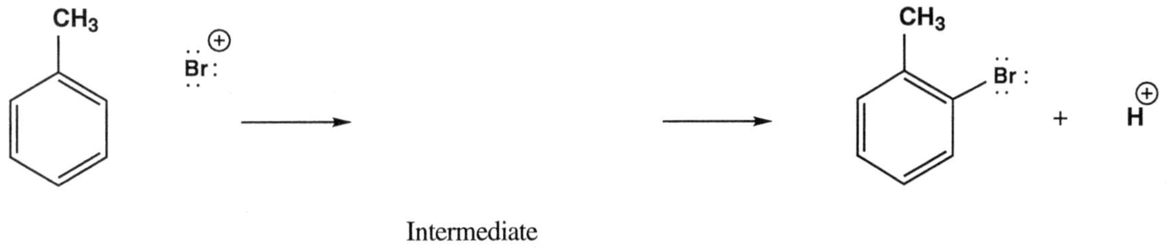
CH3
Br:
CH3
Br:
+
H
Intermediate

meta substitution

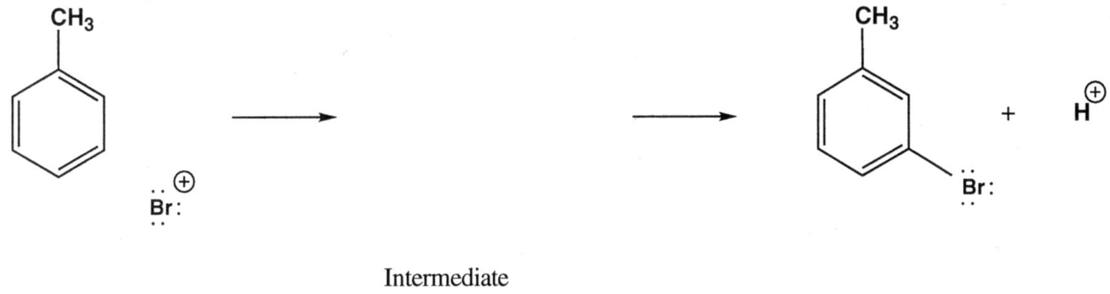
CH3
Br:
CH3
Br:
+
H
Intermediate

para substitution

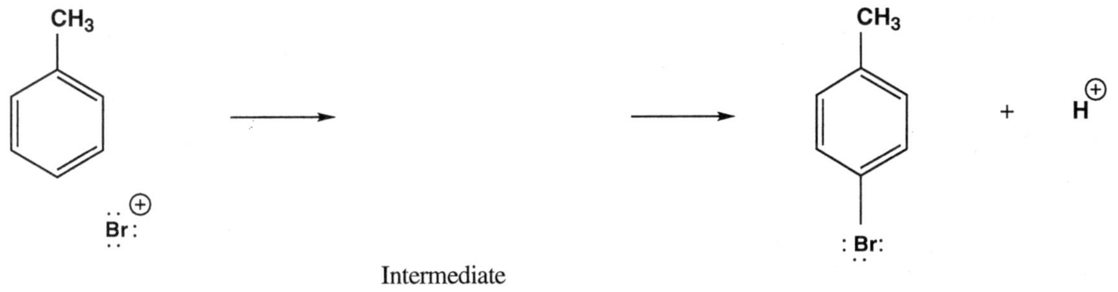
CH3
Br:
CH3
: Br:
+
H
Intermediate

Answer

ortho substitution

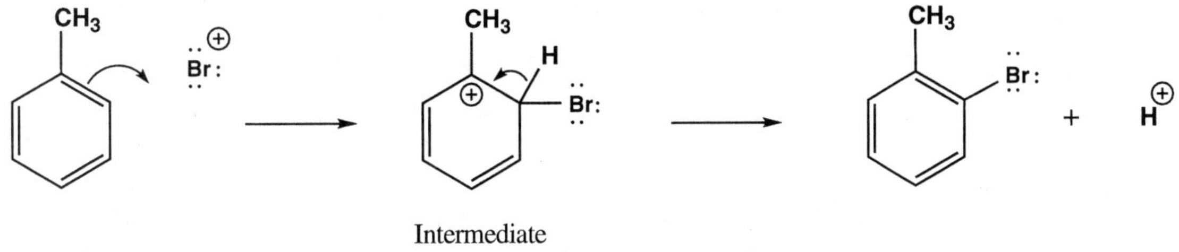
CH3
Br:
CH3
H
Br:
CH3
Br:
+
H
Intermediate

cont.

meta substitution

Intermediate

para substitution

Intermediate

7.2 The substituent effect

We have seen that the presence of a substituent complicates matters because three different products are possible, but the story is even more complicated. Of the three possible isomers which can be obtained from the bromination of toluene, two (the *ortho* and *para*) are preferred. Furthermore, the bromination of toluene goes at a faster rate than the bromination of benzene. Why?

The answer lies in the fact that the substituent already present on the ring can affect the rate and the position of further substitution. A substituent can either activate or deactive the ring towards electrophilic substitution and does so through inductive or resonance effects. Also, a substituent can direct the next substitution so that it goes mainly *ortho/para* or mainly *meta*.

We can classify substituents into four groups depending on how they affect the rate and position of substitution as follows:

- activating groups which direct *ortho/para* by inductive effects

- deactivating groups which direct *meta* by inductive effects

- activating groups which direct *ortho/para* by resonance effects

- deactivating groups which direct *ortho/para* by resonance effects.

(There are no substituents which activate the ring and direct *meta*.)

We shall look at each of these groups in turn.

7.3 Activating groups which direct *ortho/para* by inductive effects

A methyl substituent is one such group and so we shall consider again the bromination of toluene. In order to explain the directing properties of the methyl group, we need to look more closely at the mechanisms involved in generating the *ortho*, *meta* and *para* isomers.

The ease with which a particular electrophilic substitution takes place depends on the stability of the intermediate which is formed during the reaction. The more stable the intermediate, the less reactive it is. Assuming that the intermediates above have different stabilities, would you expect the reaction to go faster or slower for the less stable, more reactive intermediates?

Answer

The reaction would go slower.

That question was a bit of a trap. You may well have thought that the reaction would go faster for the more reactive intermediate. That is not the case. If an intermediate is reactive, it is less stable. Would you expect such an intermediate to be formed easily?

Answer

No. The more stable an intermediate is, the easier it can be formed during the reaction and it is that which determines whether one particular reaction pathway is followed over another. The reaction pathway preferred will be the one which goes through the most stable intermediate because that is the intermediate most easily formed.

Having settled that principle, we now have to decide if the intermediates in the three possible mechanisms above have different stabilities. Draw these intermediates and compare them. You have been told that the *ortho* and *para* products are formed, in preference to the *meta* product. Which are the more stable intermediates in that case?

Answer

| *ortho* intermediate | *meta* intermediate | *para* intermediate |

The *ortho* and the *para* intermediates must be more stable than the *meta* intermediate.

We now have to explain why the *ortho* and *para* intermediates are more stable.

The intermediates are charged and it is this charge which needs to be stabilised. In general, how can a positive charge be stabilised? (There are two general methods and you have already encountered them in previous sections.)

Answer

The positive charge can be stabilised if it is delocalised or spread round the structure by resonance. Alternatively, electron-donating groups next to the positive charge can lower the intensity of the charge and help to stabilise it.

Compare the *ortho* intermediate with the *meta* intermediate. Can you see any way in which the *ortho* intermediate can be stabilised inductively more than the *meta* intermediate? (If you are unsure, revise how carbocations are stabilised in the electrophilic addition of alkenes.)

| *ortho* intermediate | *meta* intermediate |

Answer

In the *ortho* intermediate, the positive charge can be stabilised by the inductive effect of the neighbouring methyl group. This group can push electrons towards the charged centre. This is not possible with the *meta* intermediate.

ortho intermediate
stabilised

meta intermediate

The inductive effect of the methyl group helps to explain the greater stability of the *ortho* intermediate over the *meta*. Does it explain the greater stability of the *para* intermediate?

Answer

At first sight, no. The charge on the *para* intermediate is also on an unsubstituted part of the ring. However, you must remember that the positive charge on these intermediates is not localised on one atom. It can be spread round the ring by resonance (see Section 6). This spreading of charge (delocalisation) also helps to stabilise the charge.

If necessary, revise Section 6 on the different resonance structures which are possible for the intermediate ion, then draw the resonance structures for the three intermediates above. Include the curly arrows to show how the electrons are moving about to give you the different resonance structures.

ortho intermediate

cont.

meta intermediate

para intermediate

Answer

ortho intermediate

meta intermediate

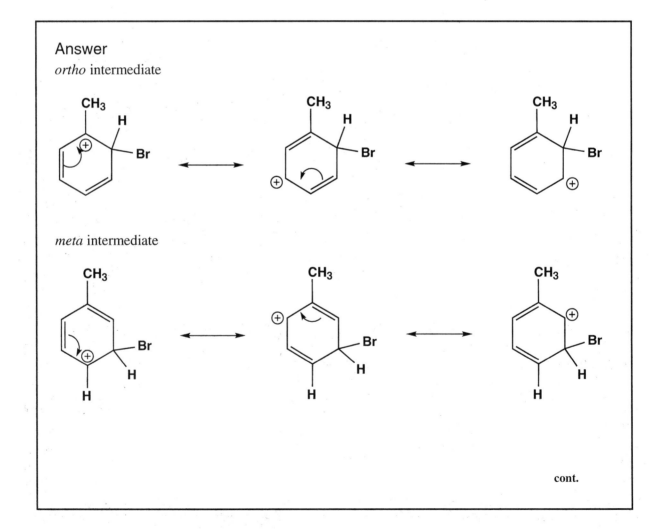

cont.

para intermediate

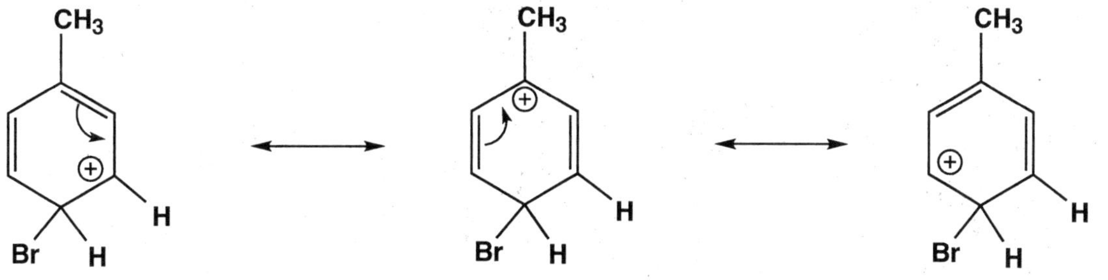

Each intermediate has three resonance structures and this helps to stabilise the charge in each intermediate. However, that by itself does not explain why the intermediate leading to *meta* substitution is less stable compared to the intermediates for *ortho* and *para* substitution. If all three intermediates have the same number of resonance structures, they should all be stabilised to the same extent.

Obviously, there is another effect which gives the *ortho* and *para* intermediate resonance structures added stability. We have already seen what that is with the *ortho* intermediate. Take a close look at the resonance structures for the *meta* and *para* intermediates and see if you can identify any which are likely to be stabilised in a similar manner. Outline the ones you think are stabilised and explain why you think they are more stable.

ortho

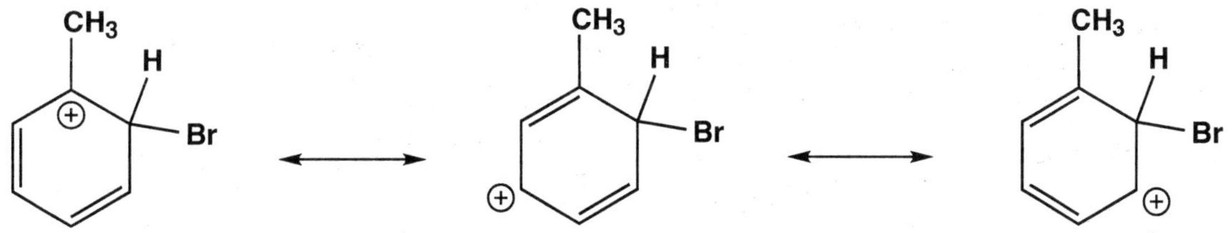

meta

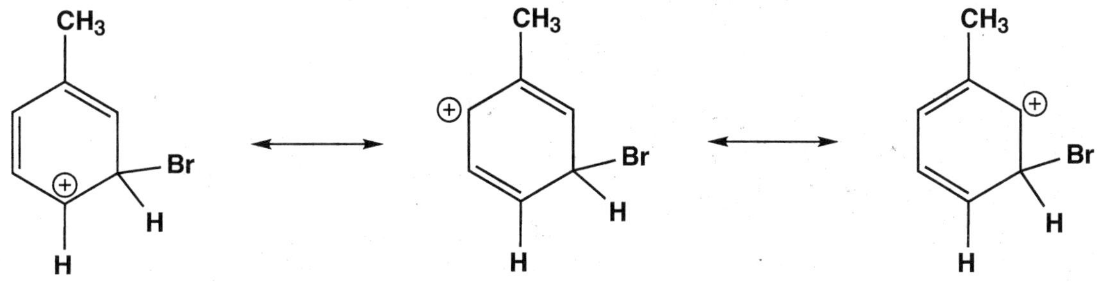

para

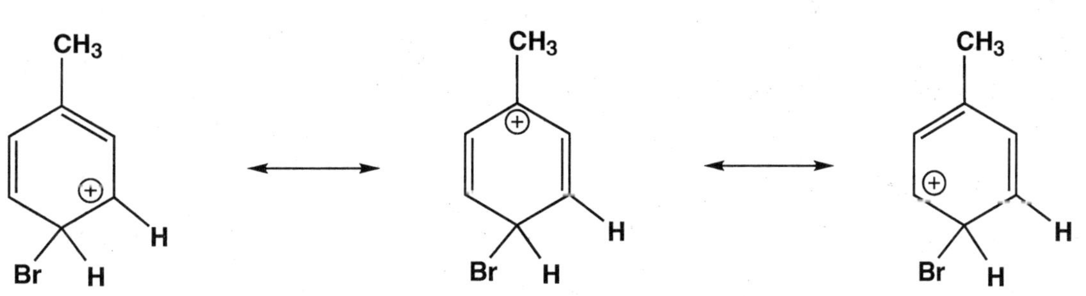

Answer

For the *ortho* and the *para* intermediates, there is one resonance structure which places the positive charge next to the methyl substituent. That resonance structure will be more stable than the other two since the methyl group can 'push' electrons towards the positive centre and lower the charge. None of the three resonance structures for the *meta* intermediate have the positive charge next to the methyl group and so none of them have this extra stabilisation.

ortho intermediate

ortho intermediate

meta intermediate

para intermediate   Stabilised

To sum up, the extra stabilisation resulting from the inductive effect of the methyl group is responsible for making the intermediates for *ortho* and *para* attack more stable than the intermediate for *meta* attack. As a result, the reaction prefers to go *ortho* or *para*.

Based on the arguments so far, would you expect toluene to be more reactive than benzene towards electrophilic substitution?

Answer

Toluene will be more reactive than benzene towards electrophilic substitution due to the extra contribution which the methyl group makes towards the stabilisation of the reaction intermediate. The electron-donating effect into the ring also increases the electron density in the ring, making it more nucleophilic and more reactive to electrophiles.

The methyl group is known as an **activating group** for electrophilic substitution (i.e. it activates the aromatic ring such that electrophilic substitution occurs more easily). The methyl group is also ***ortho*/*para* directing** since it encourages *ortho* and *para* substitution over *meta* substitution.

Predict the possible products which could be formed from the nitration of toluene and explain which products are most likely.

Answer

ortho (58%)   *meta* (4%)   *para* (38%)

Three possible products are possible for this reaction. The *ortho* and *para* isomers are more likely than the *meta* isomer since the methyl group directs *ortho*, *para*. The actual product ratios are given.

The amount of *meta* substitution is very small, as we would expect, but why is there more *ortho* substitution compared to *para* substitution? According to what we have said, there should be no great preference for *ortho* attack over *para*. The intermediates both have three resonance structures where one of the resonance structures has the extra stability afforded by the methyl group. Why then is there more *ortho* product? (Hint: The answer is based on probability.)

Answer

Quite simply, there are two *ortho* sites on the molecule to one *para* site and so there is twice as much chance of *ortho* attack as *para* attack.

Based on pure statistics we would have expected the ratio of *ortho* to *para* to be 2:1. In fact, the ratio is closer to 1.5:1. In other words, there is **less** *ortho* substitution than expected. There must be some factor which makes *ortho* substitution slightly more difficult than *para* substitution. Can you think what that might be? Think of the relative ease with which the electrophile can approach the two different sites.

Answer

Since the *ortho* sites are immediately 'next door' to the substituent, the size of the substituent will tend to interfere with *ortho* attack—a steric effect. There will be no steric effect for *para* substitution.

The methyl group is not the only group which can activate aromatic rings towards electrophilic substitution. Other alkyl groups will act as activating groups and direct *ortho* and *para*.

Give the expected products for the following reactions.

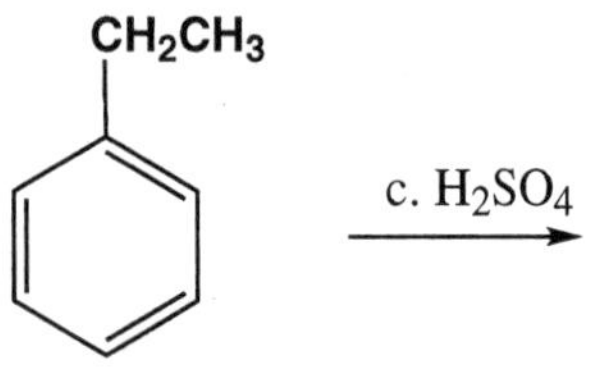

CH$_2$CH$_3$

c. H$_2$SO$_4$ →

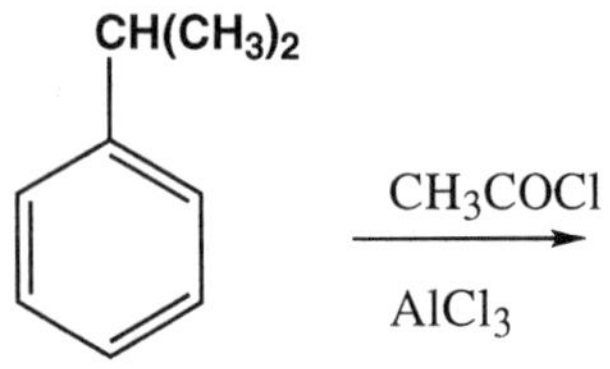

CH(CH$_3$)$_2$

CH$_3$COCl / AlCl$_3$ →

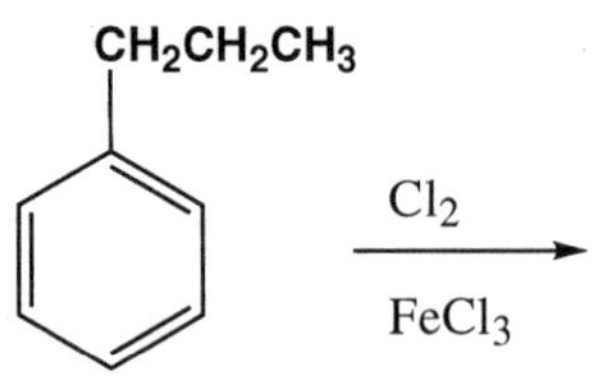

CH$_2$CH$_2$CH$_3$

Cl$_2$ / FeCl$_3$ →

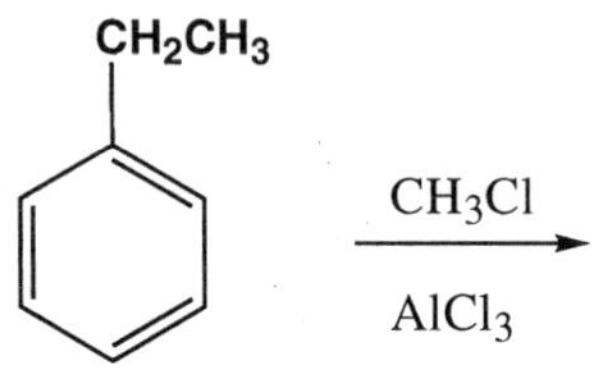

CH$_2$CH$_3$

CH$_3$Cl / AlCl$_3$ →

Answer

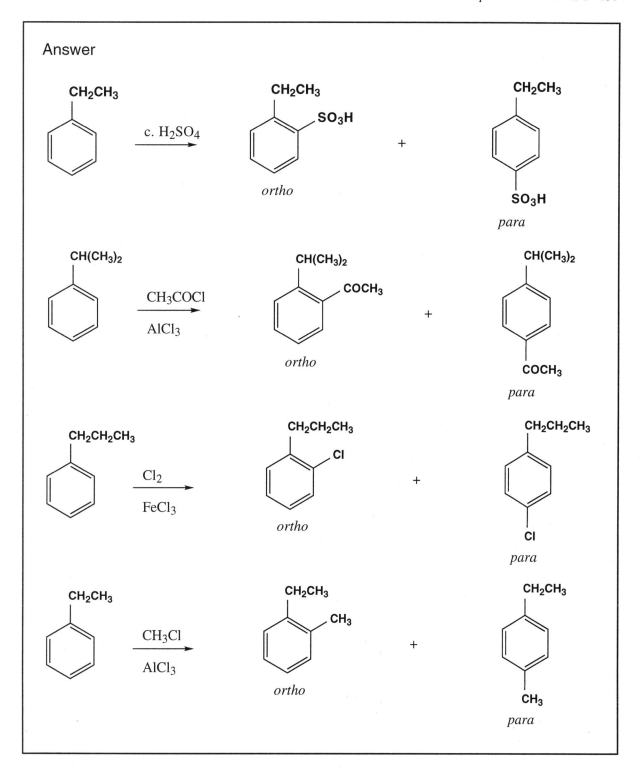

Although alkyl groups are activating and *ortho*, *para* directing, the ratio of *ortho* to *para* can vary.

Compare the following bromination reaction carried out on toluene and on 2-phenylpropane. The ratio of *ortho* product to *para* product is significantly lower for the reaction with 2-phenylpropane. Can you suggest why? Compare the two alkyl substituents.

Toluene → ortho + para

2-Phenylpropane → ortho + para

Answer

The isopropyl substituent in 2-phenylpropane is significantly bulkier than the methyl group on toluene. This has the effect of 'shielding' the *ortho* sites, making approach to these sites more difficult.

7.4 Deactivating groups which direct *meta* by inductive effects

We have seen that alkyl groups direct *ortho/para* and are activating groups. This is due to the fact that the alkyl group has an inductive effect on the ring where it 'pushes' electrons onto the ring and helps to stabilise the intermediates in the reaction.

Suppose the substituent was electron-withdrawing instead. Would you expect such a substituent to activate or deactivate the ring towards electrophilic substitution? Explain your answer in terms of the effect the substituent has on (a) the nucleophilicity of the ring and (b) the stability of the intermediate.

Answer

An electron-withdrawing group would deactivate the ring. If the group withdraws electrons from the ring, it makes the ring less nucleophilic and less likely to react with an electrophile. The electron-withdrawing effect also destabilises the reaction intermediate making the reaction more difficult.

The following groups are deactivating groups:

NO_2, SO_3H, NR_3^+, CO_2H, COR, CHO, CN

Draw the full structures of these substituents. Do they have anything in common? Look in particular at the atom which would be directly attached to the aromatic ring.

Answer

They all have a positively charged atom or an electron-deficient atom (i.e. an electrophilic centre) directly attached to the aromatic ring. Since this atom is electron-deficient, it pulls electrons towards it. This has the effect of withdrawing electron density from the ring.

Deactivating groups make electrophilic substitution more difficult but they do not make it impossible. The reaction will still go under more forcing conditions. Draw the intermediates (all possible resonance structures) which would be involved in *ortho*, *meta* and *para* attack for the bromination of nitrobenzene and decide what effect an electron-withdrawing substituent has on the stability of the intermediates. Which intermediate is most stable?

ortho

meta

para

Answer
ortho substitution

Destabilised

cont.

meta substitution

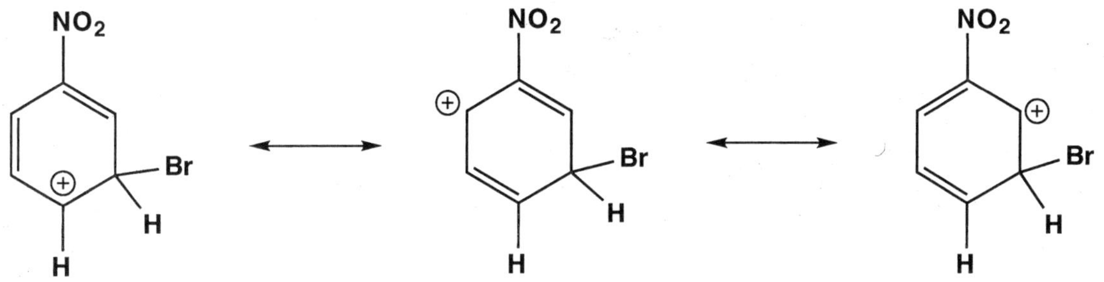

para substitution

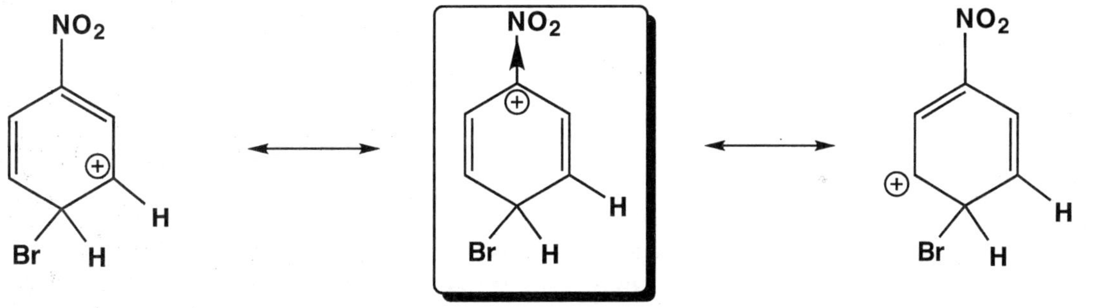

Destabilised

With *ortho* and *para* attack, one of the resonance structures has its positive charge immediately next to the substituent. Since the substituent is electron-withdrawing, this resonance structure is highly unstable and would not contribute to the stability of the reaction intermediate.

If we now look at the intermediate involved in *meta* attack, all three resonance structures are perfectly acceptable. None of them has the positive charge next to the substituent.

What product would be preferred and why?

Answer

The *meta* product will be preferred since its intermediate is more stable, having three acceptable resonance structures.

To sum up, electron-withdrawing groups, such as the nitro group, deactivate the aromatic ring towards electrophilic substitution. More forcing conditions are required for the reaction to take place and the substituent directs substitution to the *meta* position.

7.5 Activating groups which direct *ortho/para* by resonance effects

We now look at phenol. Here the substituent is a hydroxyl group. What atom is next to the aromatic ring? Is it electronegative?

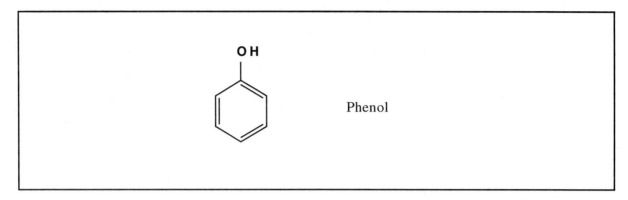

Phenol

Answer

Oxygen is next to the aromatic ring, and it is electronegative.

Since oxygen is electronegative, would you expect its inductive effect on the ring to be electron-withdrawing or electron-donating? Based on that answer, would you expect oxygen to be an activating or deactivating group?

Oxygen is electronegative and would have an electron-withdrawing inductive effect. You would expect the hydroxyl group to deactivate the ring based purely on its inductive effects.

In fact, the phenolic group is a powerful activating group. This is due to the fact that oxygen is electron-rich and can also act as a nucleophile. Let us look at a specific reaction before proceeding further.

Draw the three possible products for the nitration of phenol.

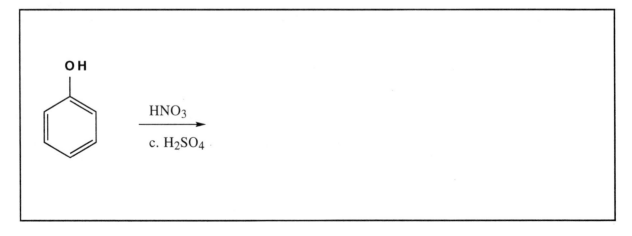

Answer

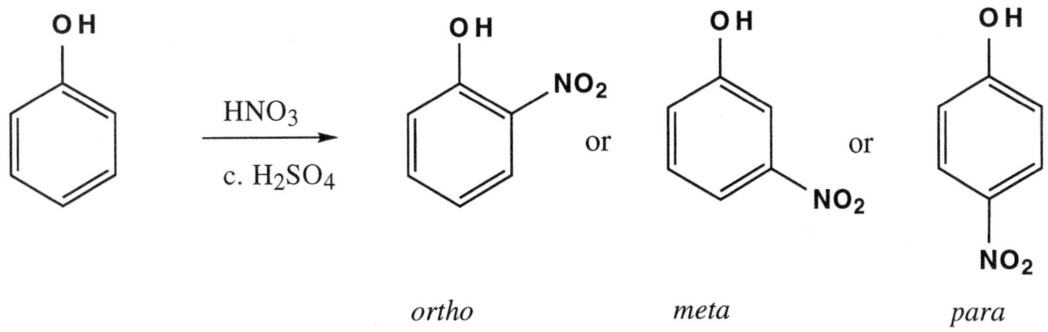

or

or

ortho *meta* *para*

Give the mechanism for the three different modes of attack.

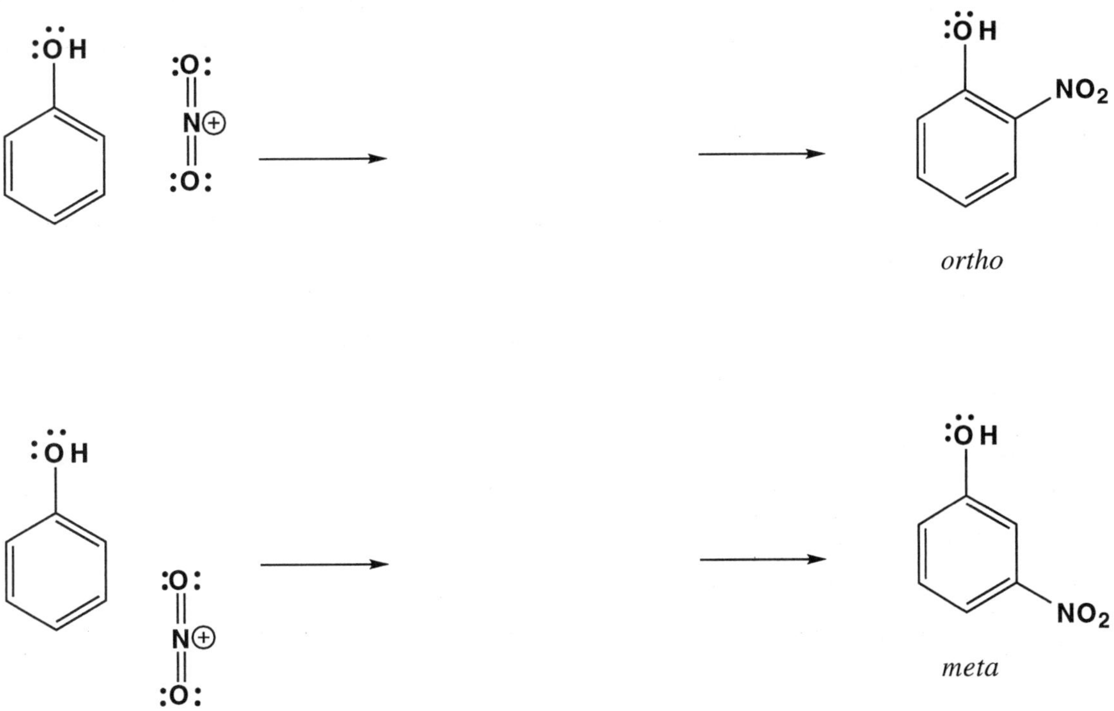

ortho

meta

Answer

ortho substitution

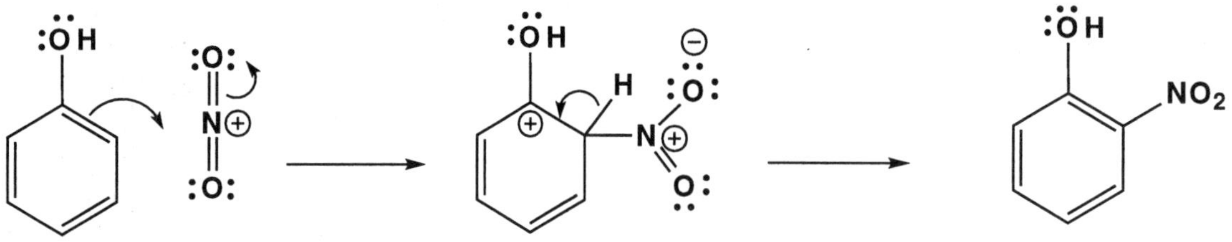

ortho

meta substitution

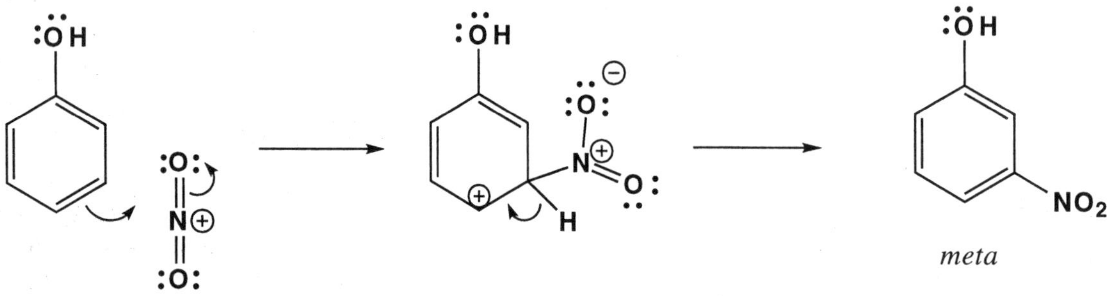

meta

para substitution

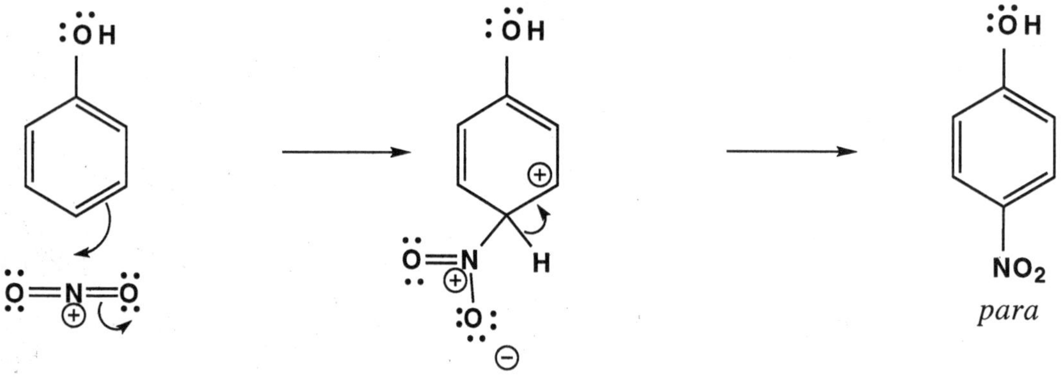

para

Draw all the possible resonance structures for the intermediates in these mechanisms. Draw arrows to show the movement of electrons.

ortho intermediate

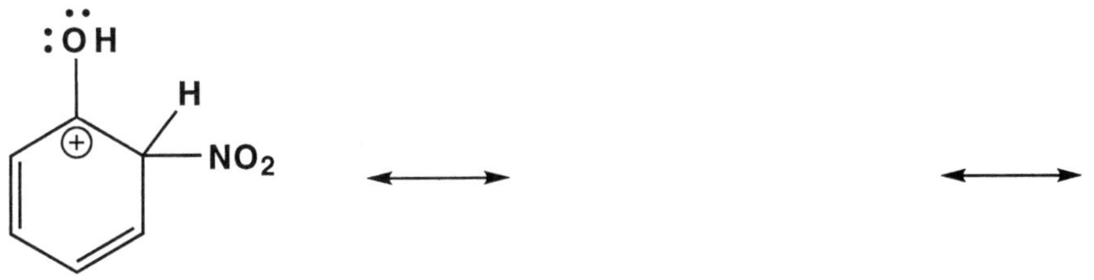

meta intermediate

para intermediate

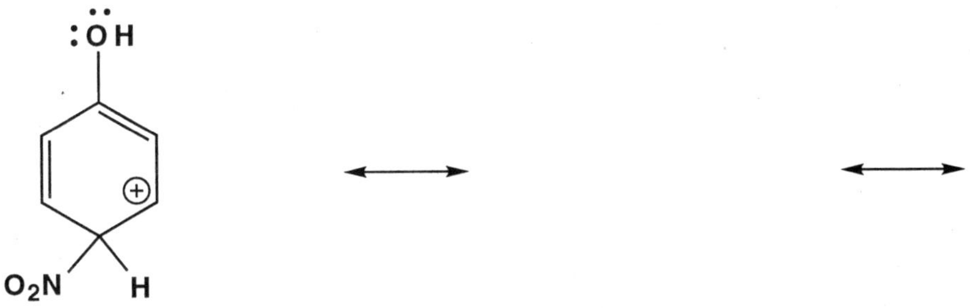

Answer

ortho intermediate

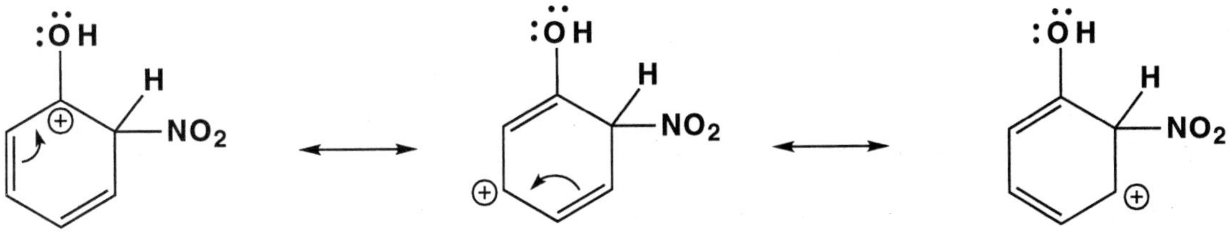

cont.

meta intermediate

para intermediate

Which resonance structures have the positive charge next to the OH substituent. Do they result from *ortho*, *meta* or *para* substitution?

Answer

The resonance structures having the positive charge next to the OH substituent are formed by *ortho* and *para* substitution.

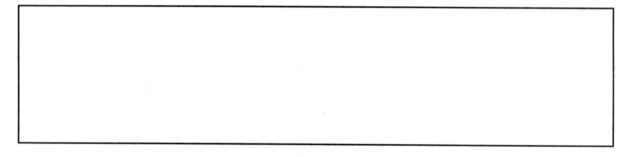

If oxygen only has an inductive effect, these resonance structures would be highly unstable. However, we know that oxygen is a powerful activating group and is stabilising the charge. This is due to the fact that oxygen can act as a nucleophile. What property does a nucleophile have?

Answer

A nucleophile is electron-rich and can form bonds to electrophiles.

Identify the nucleophilic and the electrophilic centres in the crucial resonance structures.

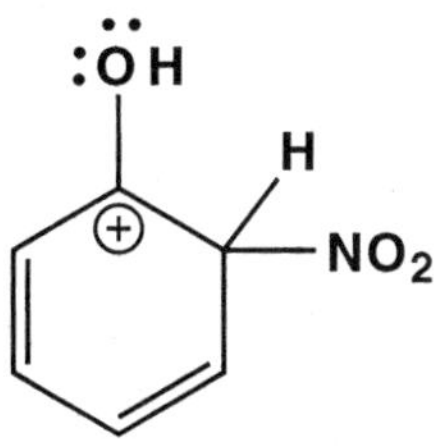

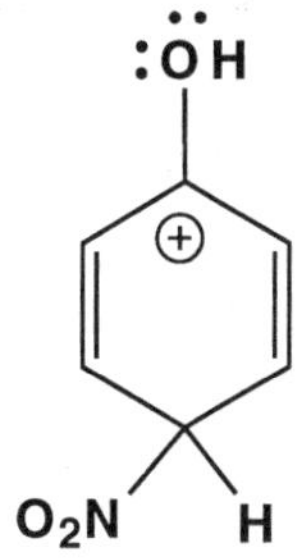

Answer

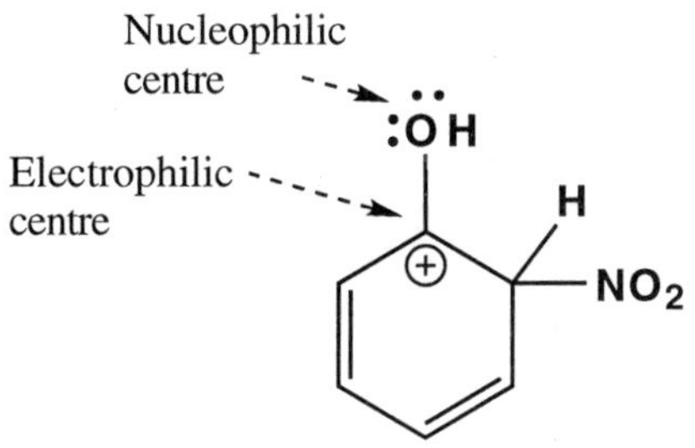

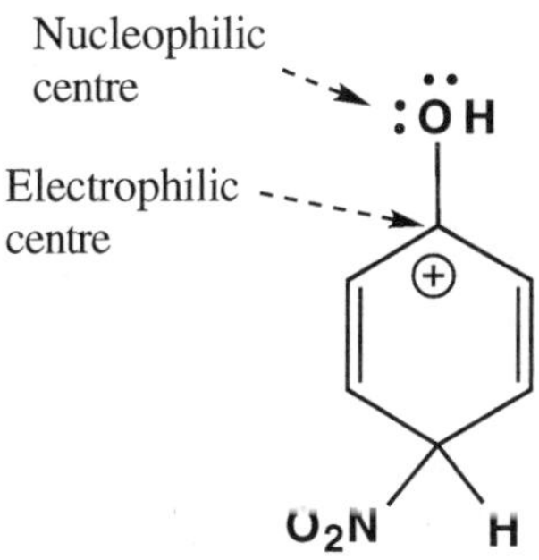

Oxygen can now use one of its lone pairs of electrons to form a bond to the neighbouring electrophilic centre. This bond will be a π bond since there already is a bond between these two atoms. For both resonance structures, draw the mechanism for this bond formation and give the structures of the new resonance structure which is formed.

Answer

What has happened to the positive charge?

Answer

Since oxygen has been able to form a new bond to the aromatic ring, it has created a fourth resonance structure which has spread the positive charge out of the ring and onto the oxygen atom.

What effect would this further delocalisation of charge have on the stability of the intermediate and the rate of reaction?

Answer

Delocalising the charge helps to stabilise it further. This in turn stabilises the intermediate, which makes the reaction go faster.

To conclude, the phenol group is an activating group which is *ortho/para* directing because of resonance effects. This resonance effect is more important than any inductive effect which the oxygen might have.

The same holds true for the following substituents:

- methoxy (OMe),
- amines (NH_2, NHR, NR_2) and
- amides (NHCOR).

In all these cases, there is either a nitrogen or an oxygen next to the ring. Both these atoms are nucleophilic and have lone pairs of electrons which can be used to form an extra bond to the ring and give a fourth resonance structure, thus spreading the positive charge out of the ring.

The ease with which the group can do this depends on how nucleophilic the atom is and how well it can cope with the positive charge it gains. Taking that into account, would you expect an NH_2 group to be a stronger activating group than a phenol?

Answer

A nitrogen atom is more nucleophilic than an oxygen atom, since it is able to cope better with the resulting positive charge. You would therefore expect amine substituents to be stronger activating groups.

Would you expect an amide group to be a stronger activating group than an amine?

Answer

You have already met this question in a different section. The amide group would be a weaker activating group since the nitrogen atom is less nucleophilic. This is because the nitrogen's lone pair of electrons is being pulled towards the carbonyl group and is less available to form a bond to the ring.

This property of amides can be quite useful. Suppose for example you wanted to make *para*-bromoaniline. What would be the most obvious way of attempting this, starting from benzene (a three-step process)?

Answer

In theory, this reaction scheme should give the right product. Benzene is nitrated and the nitro group is reduced to give aniline. The NH_2 group would activate the ring and direct the bromo group to the *ortho* and the *para* positions. The *para* product could then be separated from the *ortho* product.

In fact, the NH_2 group is such a strong activating group that bromination goes three times, such that the product obtained is the following.

How then could you adapt the above synthesis to get the product you want, bearing in mind what we have already discussed (a five-step process)?

Answer

Since the NH_2 group is too strong as an activating group, we can convert it to an amide which is less activating. The reaction then only goes once. We also find that the reaction is more selective for the *para* position than for the *ortho* position. Why is this?

Answer

The amide group is bulkier than the NH_2 group and tends to shield the *ortho* positions from attack.

We have stated that the amino group (-NH_2) is *ortho / para* directing. However, the nitration of aniline gives the meta product preferentially. Why is this apparently strange result obtained? (Hint: The reaction takes place under acid conditions. See also page 139.)

Preferred Product

Answer

Under acid conditions, the amino group is protonated to give the ammonium ion. The lone pair is now 'tied up' in a bond, and so resonance into the ring is no longer possible. The -NH_3^+ group is a deactivating group which withdraws electron density from the aromatic ring by an inductive effect. This will destabilise the intermediates for *ortho* and *para* substitution. Therefore, *meta* substitution is preferred.

7.6 *ortho/para* directing groups which deactivate the ring

So far we have identified three types of aromatic substituent.

* *ortho /para* directing groups which activate the ring by inductive effects (alkyl groups).

* *meta* directing groups which deactivate the ring by inductive effects (NO_2, SO_3H, CN, COR, NR_3^+, CO_2H, CHO).

* *ortho /para* directing groups which activate the ring by resonance (OH, NH_2, NHR, NR_2, NHCOR).

The fourth and last group of aromatic substituents which we are going to discuss are the halides. These are perhaps the trickiest to understand since they deactivate the ring by one effect, but direct substitution by a different effect.

Let us tackle the deactivation of the ring first of all. Is a halogen atom electronegative? If so, would you expect it to activate or deactivate the aromatic ring to electrophilic substitution?

Answer

The halide ion is strongly electronegative and therefore we would expect it to have a strong electron-withdrawing inductive effect on the aromatic ring. This would make the aromatic ring less nucleophilic and less reactive to electrophiles. It would also destabilise the required intermediate for electrophilic substitution.

That makes sense and fits in with the fact that halide substituents deactivate the ring. We noted above that oxygen and nitrogen are also electronegative but that this is less important than the resonance effect which activates the ring. What conclusion can you draw regarding the relative importance of inductive and resonance effects for halide substituents?

Answer

Since halide substituents deactivate the ring, this must imply that the inductive effect is more important than the resonance effect, resulting in overall deactivation of the ring.

Why should the inductive effect be stronger for halide ions compared to oxygen or nitrogen?

Answer

Halide substituents should have a stronger electron-withdrawing effect since they are more electronegative than either oxygen or nitrogen. They are also poorer nucleophiles and so resonance effects are less important.

So far, everything makes sense, but now comes the tricky bit. If halide ions are deactivating the ring by electron-withdrawing effects, then why do they not direct substitution to the *meta* position like other electron-withdrawing groups (e.g. the nitro group)?

Let us look at a specific reaction—the nitration of bromobenzene. Draw the three possible products for this reaction.

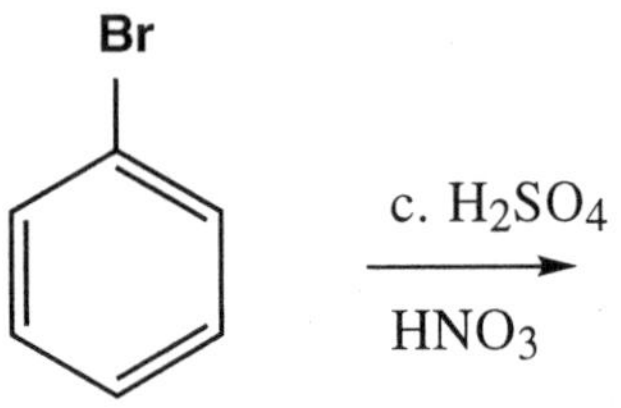

Answer

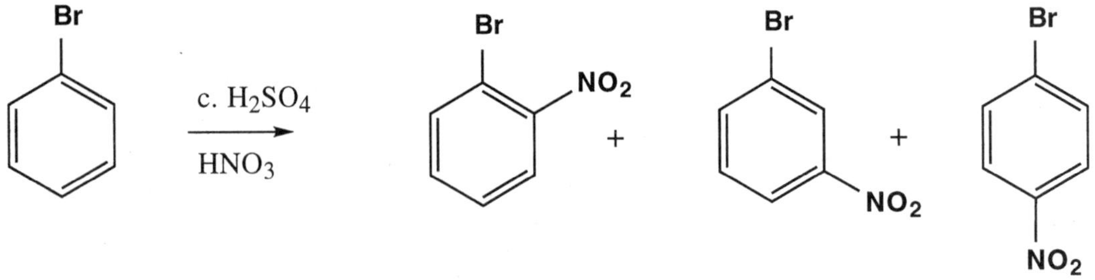

ortho   meta   para

Draw the mechanisms leading to these three different products.

ortho substitution

meta substitution

para substitution

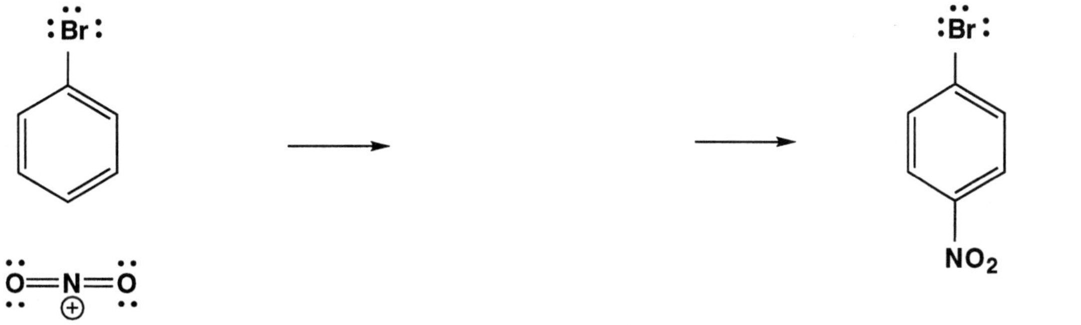

Answer

ortho substitution

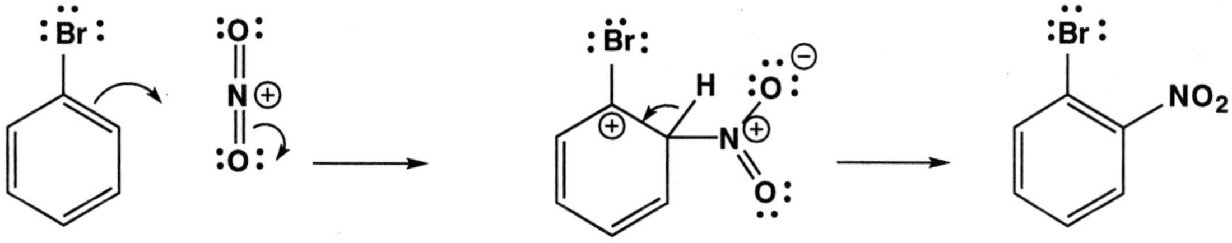

cont.

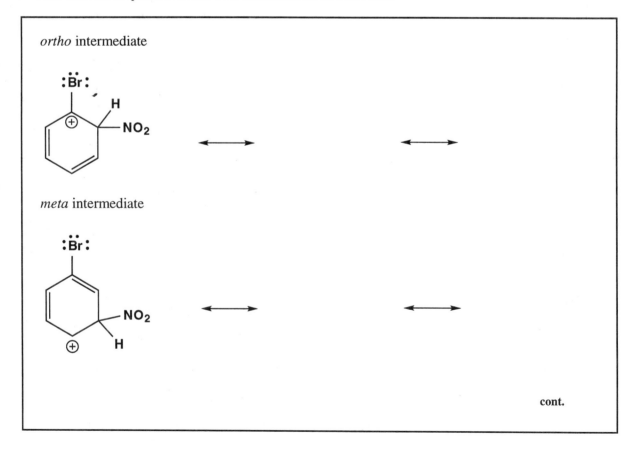

meta substitution

para substitution

Now draw all the possible resonance structures for the intermediates.

ortho intermediate

meta intermediate

cont.

para intermediate

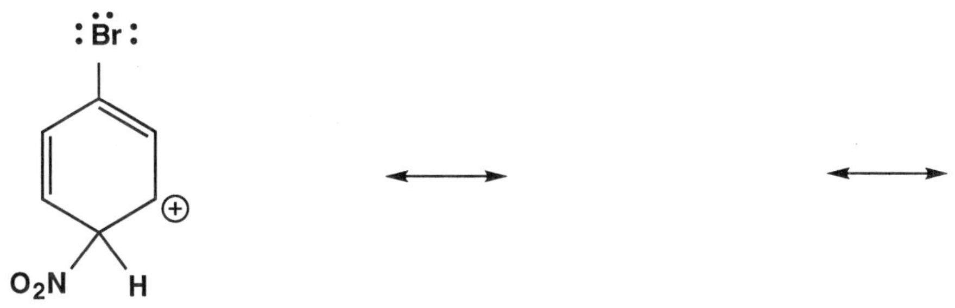

Answer

ortho intermediate

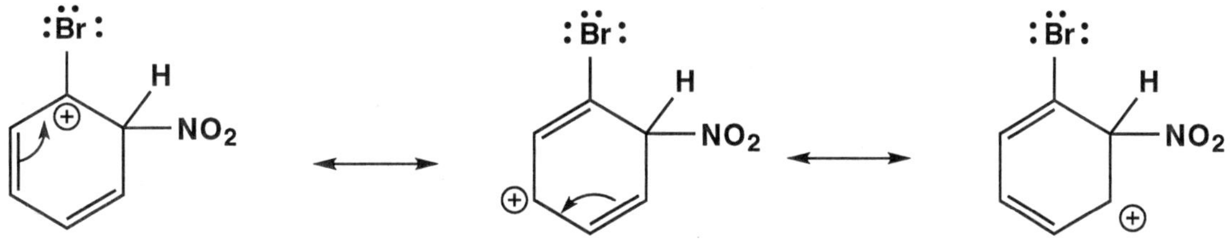

meta intermediate

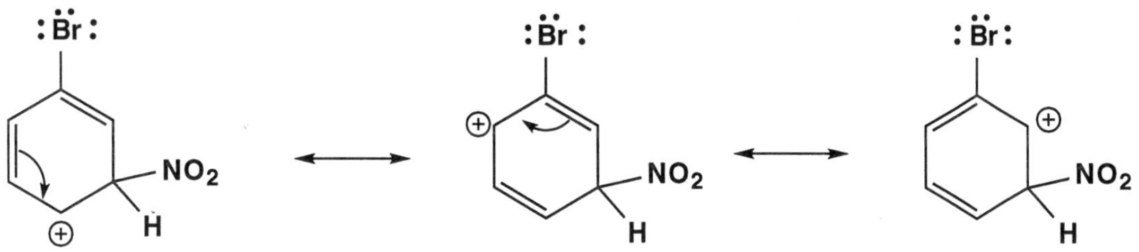

para intermediate

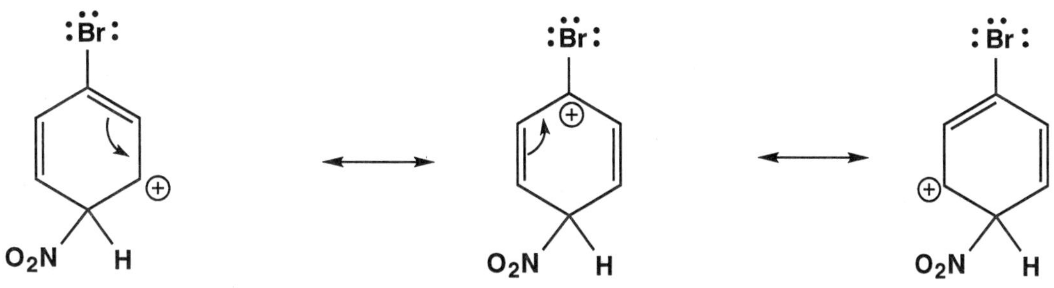

Now identify the resonance structures which have the positive charge next to the halide substituent.

Answer

ortho attack

para attack

These are the crucial resonance structures as far as the directing properties of the substituent are concerned. If bromine acts inductively, it will destabilise these intermediates and direct substitution to the *meta* position. However, we know that bromine directs *ortho/para* and so must be stabilising the *ortho/para* intermediates rather than destablising them.

How could a bromine substituent stabilise a neighbouring positive charge? (Compare what happens with N or O substituents.) Show mechanisms for the stabilisation.

Answer

The bromine substituent can stabilise the neighbouring positive charge by a resonance effect where the bromine acts as a nucleophile and donates one of its lone pairs to form a new bond to the electrophilic centre beside it. A new π bond is formed and the positive charge is now on the bromine ion.

Would you expect this resonance to be as strong for a halide substituent as for an oxygen or nitrogen substituent? Explain.

Answer

The resonance effect will be weaker in the case of a halogen. This is because the halogen is a weaker nucleophile than oxygen or nitrogen and less capable of stabilising a positive charge.

You now have all the facts. Can you come up with an explanation as to why halogen substituents deactivate the ring, yet direct *ortho* /*para,* whereas O and N substituents activate and direct *ortho* /*para*?

Answer

It is clear that the inductive and the resonance effects for oxygen, nitrogen and halogens are working in opposite directions. The inductive effect deactivates the ring, whereas the resonance effect activates the ring.

With nitrogen and oxygen substituents, the resonance effect is more important than the inductive effect and so these groups are *ortho/para* directing and activate the ring. With halogens the opposing effects are about equal. The inductive effect is strong enough to deactivate the ring, but once a reaction does take place, the resonance effect predominates over the inductive effect and forces the reaction to go *ortho* or *para*.

Fluoro, bromo, chloro and iodo substituents all deactivate aromatic rings towards electrophilic substitution by an electron-withdrawing inductive effect. However, they direct substitution to go *ortho/para* by a resonance effect. With opposing forces such as this, the balance between the factors is crucial to the net result.

7.7 Planning a synthesis

A full understanding of how substituents direct further substitution is crucial if you are planning the synthesis of a multi-substituted aromatic compound. For example, how would you synthesise the following aromatic compound from benzene?

You have two choices. You can either brominate first, then nitrate. or you can nitrate first then brominate. Which would you try? Show what products you would expect from each route.

Answer

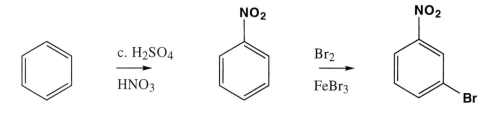

Nitrating first then brominating second would give the *meta* isomer and very little of the *para* isomer due to the *meta* directing properties of a nitro group.

The second method is better since the directing properties of bromine are in your favour. You would have to separate the *para* product from the *ortho* product, but you would still get a higher yield by this route.

How would you make the following disubstituted aromatic rings starting from benzene?

(a)

(b)

(c)

(a)

(b)

cont.

(c)

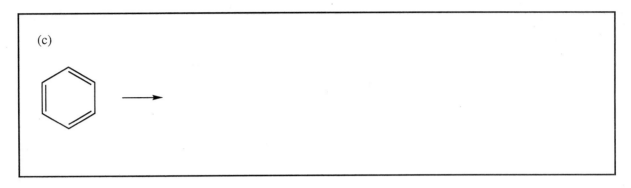

Answer

Each of these compounds has two substituents, so which substituent do you put in first? Sometimes, this does not matter if the directing properties of both substituents are the same. In these cases, however, the substituents have different directing properties, so it is important to put the correct substituent in first. The correct answers are shown below.

(a)

$$\text{benzene} \xrightarrow{\text{c. } H_2SO_4} \text{benzenesulfonic acid (SO}_3\text{H)} \xrightarrow[\text{FeBr}_3]{\text{Br}_2} \text{3-bromobenzenesulfonic acid (Br, SO}_3\text{H)}$$

(b)

$$\text{benzene} \xrightarrow[\text{AlCl}_3]{\text{CH}_3\text{Cl}} \text{toluene (CH}_3\text{)} \xrightarrow[\text{c. } H_2SO_4]{\text{HNO}_3} \text{4-nitrotoluene (NO}_2, \text{CH}_3\text{)}$$

(c)

$$\text{benzene} \xrightarrow[\text{c. } H_2SO_4]{\text{HNO}_3} \text{nitrobenzene (NO}_2\text{)} \xrightarrow[\text{FeCl}_3]{\text{Cl}_2} \text{3-chloronitrobenzene (NO}_2, \text{Cl}_)}$$

How would you make the following from benzene?

$$\text{NH}_2 \quad \text{CH}_3$$

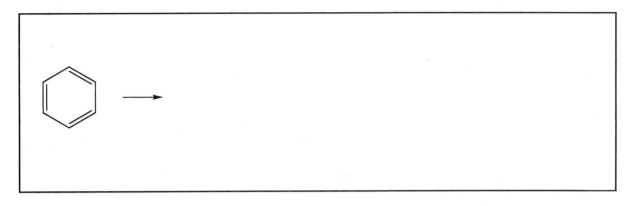

Answer

This question requires a little more thought. Both groups are activating groups and direct *ortho/para*. However, the two groups in the product are *meta* with respect to each other. In order to get *meta* substitution we have to have a deactivating group present as the first substituent.

Therefore, we need to add a deactivating group first, before adding the second group. We then need to convert the deactivating group to the correct substituent.

The nitro group is ideal since it directs *meta* and can then be converted to the required amino group.

7.8 Removing substituents

It is sometimes useful to have a substituent present to direct or to block a particular substitution, and then to remove it once the desired substituents have been added.

The following reactions are used to remove substituents from aromatic rings.

1,3,5-Tribromobenzene cannot be made directly from benzene by bromination. Why not?

Answer

Benzene can be brominated to bromobenzene. However, if bromobenzene is brominated, the next bromine will be directed to the *ortho* or the *para* position. The structure we want has the bromine substituents all *meta* to each other.

Suggest how you might synthesise 1,3,5-tribromobenzene from benzene by a four-step synthesis, with the aid of a removable substituent.

Answer

The synthesis starts by introducing an amine substituent. This strongly activates the group, allowing three bromine substituents to be added at the correct positions relative to each other. The amino group can then be removed to give the desired product.

ortho-Bromotoluene could conceivably be synthesised by bromination of toluene or by Friedel–Crafts alkylation of bromobenzene. Draw the reaction sequences for both these approaches and point out any disadvantages to these routes.

Answer

Both approaches could conceivably give the desired product since both groups direct *ortho*. However, the reaction could also go *para* and this is more likely if the electrophile is hindered from approaching the *ortho* site by unfavourable steric interactions.

An alternative strategy would be to deliberately substitute a group at the *para* position before carrying out the bromination. This group would then act as a blocking group at the *para* position and would force the bromination to take place *ortho* to the methyl group. If the blocking group could then be removed, the desired product would be obtained.

You have seen that NH_2 and SO_3H groups can be removed from an aromatic ring and so one of these groups would be a likely blocking group. Consider which group could be added and removed more easily, then suggest an alternative four-step synthesis to the desired product.

Answer

Note that both the methyl group and SO3H group direct bromination to the same position.

An alternative route would have been to add a nitro group to toluene, then to carry out the bromination. However, removing the nitro group would have involved several steps (reduction to the amino group, conversion of the amino group to the diazonium salt and hydrolysis of the diazonium salt). Removal of the SO$_3$H group is much simpler.

Summary

- The presence of a substituent affects the rate of electrophilic substitution and also has a directing effect.

- Alkyl groups are activating groups which direct *ortho* and *para* by inductive effects.

- Groups having an **electrophilic** centre directly attached to the aromatic ring are deactivating groups and direct *meta* by inductive effects (NO$_2$, NR$_3^+$, COR, CHO, CO$_2$H, SO$_3$H, CN).

- Groups having a **nucleophilic** oxygen or nitrogen atom directly attached to the ring are good activating groups which direct *ortho* and *para* by resonance effects.

- Halogens deactivate aromatic rings by inductive effects, but direct *ortho* and *para* by resonance effects.

- The order in which electrophilic substitutions is carried out is crucial to obtaining the desired substitution pattern.

- NH$_2$ and SO$_3$H groups can be used as directing groups or blocking groups and then removed from the ring.

SECTION 8

Nucleophilic Additions

Nucleophilic Additions

8.1 Nucleophilic and electrophilic centres in aldehydes and ketones

Nucleophilic addition reactions occur with aldehydes and ketones. Both these functional groups have a carbonyl functional group (i.e. C=O).

$:\!O\!:$ (double bond) C with R and R'	$:\!O\!:$ (double bond) C with R and H
Ketone (alkanone)	**Aldehyde (alkanal)**

> R = alkyl group
> R' = alkyl group which may or not be the same as **R**

As the name of the reaction suggests, the reaction involves the addition of a nucleophile to the ketone or aldehyde. Before we look at how this happens, let us study the electronic properties of ketones and aldehydes more closely. Consider propanal below. Can you identify any electronegative atoms?

$:\!O\!:\ \delta^-$
‖
CH_3CH_2—C—H

Answer

You should have identified the oxygen as being electronegative since it is to the right of carbon on the periodic table.

The electronegative oxygen is bonded to a carbon which is less electronegative. What effect do you think this might have on the electrons in the double bond?

Answer

Since the oxygen is more electronegative than carbon, it is 'hungrier' for electrons and will tend to pull the electrons in the C=O bond towards it. As a result, the electrons in the double bond will spend more of their time closer to the oxygen than to the carbon.

If the electrons are unequally shared between oxygen and carbon, what effect will this have on the polarity of the bond?

Answer

Since the electrons in the double bond spend more of their time closer to oxygen than to carbon, the C=O group is polarised such that the oxygen is slightly negative and the carbon is slightly positive.

$$:\!O\!:\delta-$$
$$\|$$
$$C\,\delta+$$

CH$_3$CH$_2$ H

Now that we have studied the electronic properties and polarity of the carbonyl bond, can you identify any nucleophilic or electrophilic centres in the molecule?

$$:\!O\!:\delta-$$
$$\|$$
$$C\,\delta+$$

CH$_3$CH$_2$ H

Answer

A nucleophilic centre is electron-rich. Therefore, the oxygen atom is the nucleophilic centre. An electrophilic centre is electron-deficient. The carbon of the carbonyl group is therefore an electrophilic centre.

Nucleophilic centre ----→ $:\!O\!:\delta$

Electrophilic centre ←--- $C\,\delta+$

CH$_3$CH$_2$ H

The following structures are aldehydes and ketones. Identify which are aldehydes and which are ketones, name them, then identify any nucleophilic or electrophilic centres present in these molecules.

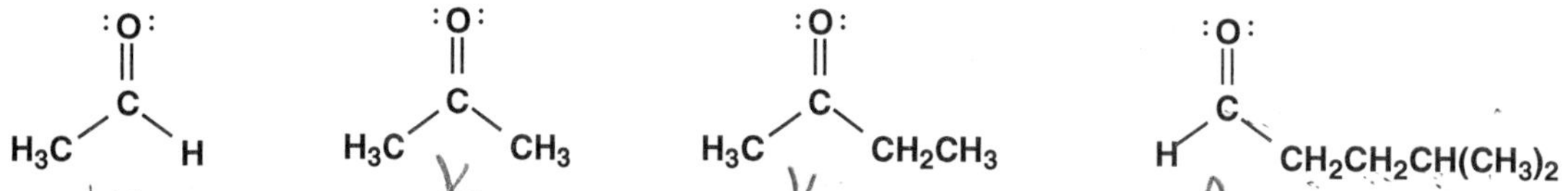

[Handwritten annotations on the four structures:]

Structure 1 (H₃C–CHO): Key, Alde, ethanal

Structure 2 (H₃C–CO–CH₃): K, propanone

Structure 3 (H₃C–CO–CH₂CH₃): K, 2-butanone

Structure 4 (HCO–CH₂CH₂CH(CH₃)₂): A, 4-methyl pentanal

Answer

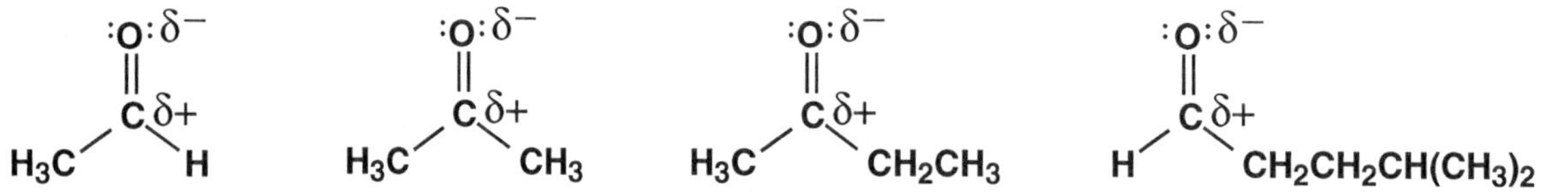

| Aldehyde | Ketone | Ketone | Aldehyde |
| Ethanal (acetaldehyde) | Propanone (acetone) | Butanone | 4-Methylpentanal |

In all these structures, the carbonyl group is polarised. The oxygen is the nucleophilic centre. The carbon is the electrophilic centre.

If a nucleophile reacts with propanone, where will it react?

Answer

A nucleophile is electron-rich and will react with an electron-poor or electrophilic centre. The electrophilic centre in propanone is the carbonyl carbon.

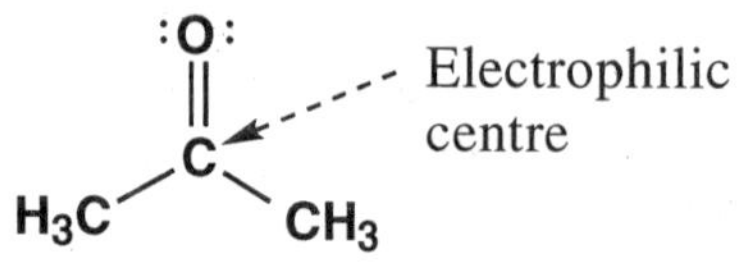

Electrophilic centre

8.2 Nucleophilic addition of cyanide ion to aldehydes and ketones

One example of a nucleophile is the cyanide ion (CN^-). We shall investigate how the cyanide ion reacts with propanone. First of all, draw the full structure of the cyanide ion and identify whether the carbon or the nitrogen atom is the nucleophilic centre.

Answer

The nucleophilic centre of the nitrile group is the carbon atom since this is the atom with the negative charge.

$$:C\!\equiv\!\!N:$$

We have now identified the nucleophilic centre of the nitrile group and the electrophilic centre of the carbonyl group. Reaction between the nitrile and the carbonyl group will lead to a bond between these two centres. If such a bond formed and nothing else happened, the carbonyl carbon would end up with five bonds. This is not possible, so one of the original four bonds to carbon must break at the same time as the new one forms. Which bond would be the most likely one to break?

:O:

C

H₃C   CH₃

New bond forming

Answer

The most likely bond to break will be the weakest one. This will be the π bond of the carbonyl bond.

If the π bond is going to break where will the two electrons making up that bond go?

Answer

They will move to the oxygen to form a lone pair of electrons on the oxygen.

Write a mechanism to show what happens when the nitrile group attacks propanone and show the structure which is formed.

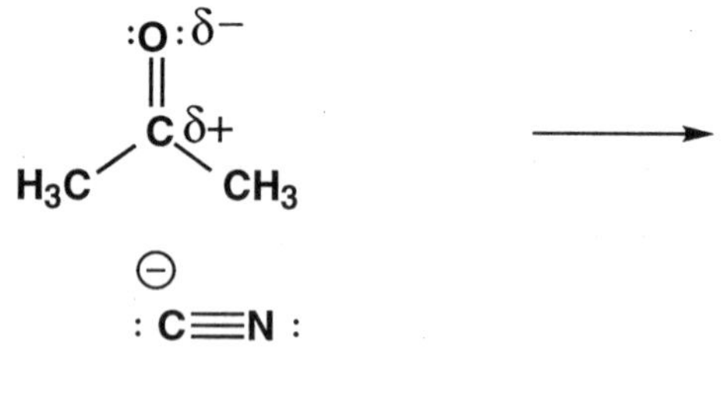

Answer

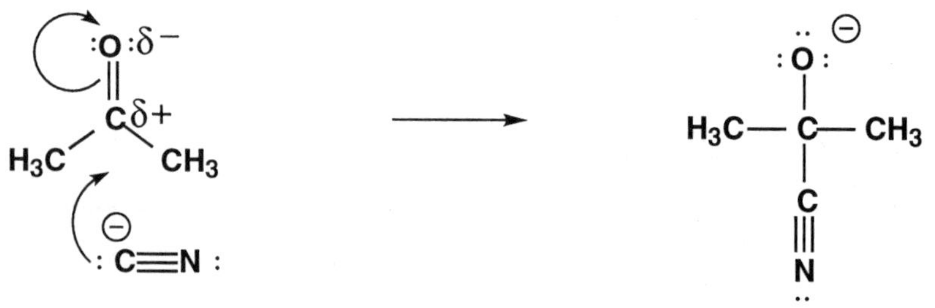

The carbonyl carbon is electrophilic due to the electron-pulling effect of the oxygen. The lone pair of electrons on the cyanide carbon atom are used to make a new bond to the carbonyl carbon. At the same time, the π bond of the carbonyl group breaks, with both electrons moving onto the oxygen to form a third lone pair. As a result, the oxygen gains a negative charge.

Note that the arrow from the nucleophile points to where the centre of the new bond will be formed. In some textbooks, you may see the arrow pointing directly to the carbonyl carbon. Formally, this is incorrect, since it implies that the electrons will end up as a lone pair on that carbon.

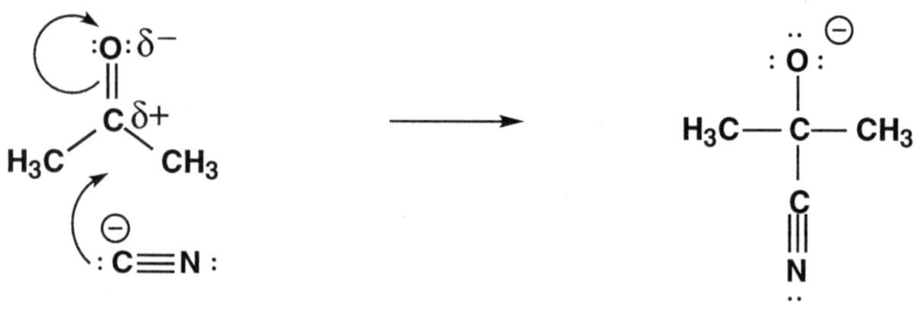

Correct

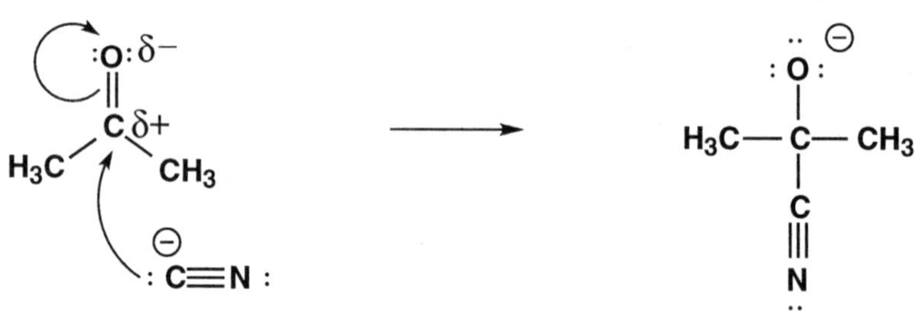

Incorrect

To carry out this reaction, propanone is treated with potassium cyanide. The potassium ion acts as the counterion both for the cyanide ion and for the reaction product.

The reaction is 'worked up' with water to protonate $O^{\ominus}$ and to give the final product—a cyanohydrin.

Draw a mechanism to show how this final product is formed from the intermediate.

Answer

8.3 Nucleophilic addition of carbanions to aldehydes and ketones

Aldehydes and ketones react with reagents called **Grignard** reagents by the same mechanism. One commonly used Grignard reagent is methylmagnesium iodide (CH_3MgI). This reagent provides the equivalent of a methyl carbanion.

In reality, the methyl carbanion is never present as an isolated ion, but the reaction proceeds as if it were. We shall therefore work on the assumption that it is present. Predict the product you would obtain if propanal was treated with methylmagnesium iodide, then aqueous acid.

Answer

The Grignard reagent provides the equivalent of a methyl carbanion, where the carbon bears a lone pair of electrons and a negative charge. This is a nucleophile and reacts with carbonyl groups in the same way as a nitrile group.

Assuming the presence of the methyl carbanion, complete the following

• identify the nucleophilic and electrophilic centres in the reagents,

• give the mechanism of the reaction,

• give an explanation for the mechanism.

Electrophilic and nucleophilic centres

cont.

Mechanism

:O:
‖
C
CH₃CH₂   H

        ⊖   H
         ╲ ╱
      : C—H
         │
         H

Explanation

Answer

The nucleophile in the reaction is the methyl carbanion. The nucleophilic centre is the carbon atom which has the lone pair of electrons and the negative charge. The electrophile in the reaction is the aldehyde. The electrophilic centre is the carbon atom of the carbonyl group since the neighbouring electronegative oxygen has a greater share of the electrons in the C=O bond.

:O:
‖
C
CH₃CH₂   H

        Electrophilic
        centre

        ⊖   H
         ╲ ╱
      : C—H
         │
         H

        Nucleophilic
        centre

Mechanism

Stage 1
The nucleophilic centre of the nucleophile forms a bond to the electrophilic centre of the electrophile. Both electrons for the bond are provided by the nucleophile (the lone pair). As the new bond is formed, a bond must break in the aldehyde since the carbonyl carbon cannot have five bonds. The weakest bond is the π bond of the carbonyl group and this is the one to break. Both electrons move onto oxygen to provide a third lone pair and a negative charge.

cont.

Stage 2
A lone pair of electrons on the oxygen is used to form a bond to a proton when the reaction product is treated with aqueous acid.

The reaction of aldehydes with Grignard reagents is a commonly used method of synthesising secondary alcohols.

Grignard reagents also react with ketones. Predict what product you would get if propanone is treated with methylmagnesium iodide and identify what kind of functional group can be synthesised when ketones are treated with Grignard reagents.

Answer

Ketones react with Grignard reagents to give tertiary alcohols.

We have seen how ketones and aldehydes react with Grignard reagents to give tertiary and secondary alcohols respectively. Can you think of a way of synthesising primary alcohols using Grignard reagents?

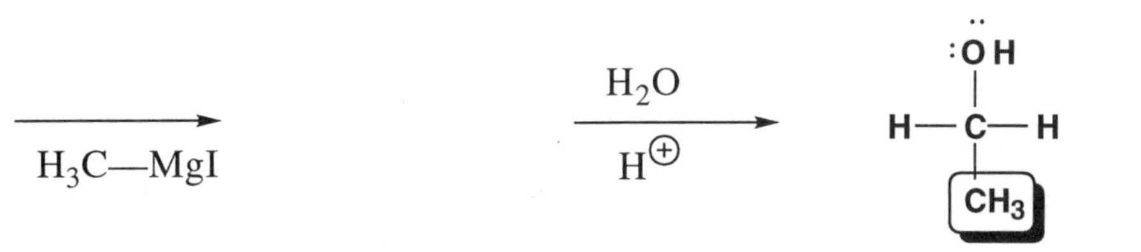

$$\xrightarrow{\hspace{3cm}} \quad H_3C\text{—MgI} \qquad \xrightarrow[H^{\oplus}]{H_2O}$$

Primary alcohol

Answer

If the Grignard reagent is treated with methanal (formaldehyde), a primary alcohol will be obtained. Example:

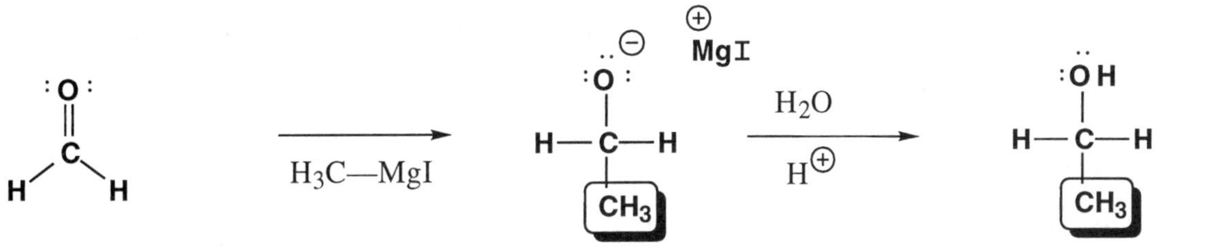

Primary alcohol

We have seen that aldehydes and ketones react with methylmagnesium iodide to give alcohols. A carbon–carbon bond is formed during the reaction, so this is an important method of building up molecules from simple starting materials. There are a large number of Grignard reagents and they can be synthesised by reacting an alkyl halide with magnesium. The magnesium is 'inserted' between the carbon and the halogen.

$$\textbf{R—X} \quad \xrightarrow[\text{diethyl ether}]{\text{Mg}} \quad \textbf{R—Mg—X} \qquad \begin{array}{l} \text{R = an alkyl group} \\ \text{X = halogen (I, Br, Cl)} \end{array}$$

Identify the Grignard reagents which were used to carry out the following reactions.

(a)

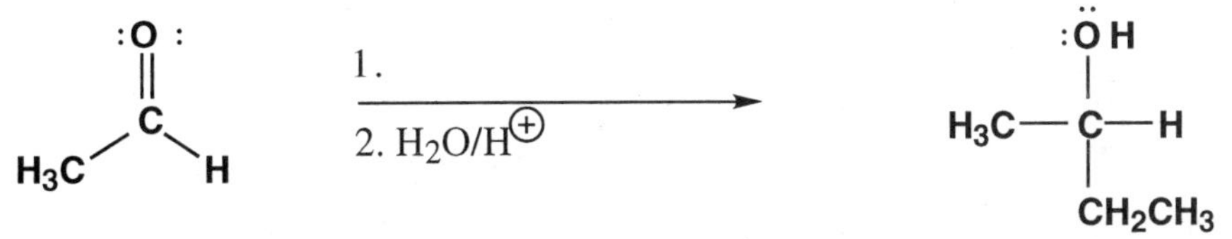

$$\xrightarrow[\text{2. H}_2\text{O/H}^{\oplus}]{\text{1.}}$$

(b)

$$\xrightarrow[\text{2. H}_2\text{O/H}^{\oplus}]{\text{1.}}$$

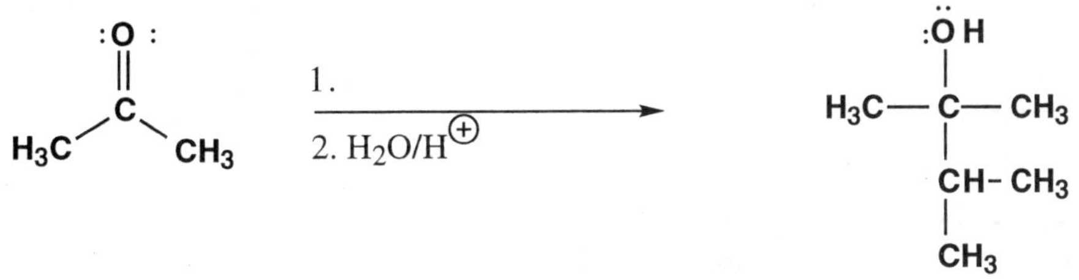

cont.

(c)

$$CH_3CH_2 \overset{\displaystyle :O:}{\underset{\displaystyle }{\overset{\displaystyle \|}{C}}} H \quad \xrightarrow[\text{2. } H_2O/H^{\oplus}]{\text{1.}} \quad CH_3CH_2 \overset{\displaystyle :\ddot{O}H}{\underset{\displaystyle CH_3}{\overset{\displaystyle |}{C}}} H$$

Answer

The first thing to do is to identify the nucleophile which has been added. Since it is a Grignard reagent which has been used, there should be an extra alkyl group.

$$\underset{R}{\overset{O}{\underset{}{\overset{\|}{C}}}}R' \quad \xrightarrow[\text{2. } H_2O/H^{\oplus}]{\text{1. } R''MgI} \quad R \overset{OH}{\underset{\boxed{R''}}{\overset{|}{C}}} R'$$

$\boxed{R''}$ - - - - Extra alkyl group

The alkyl groups on the ketone or aldehyde will still be there, so identify these first. The remaining alkyl group will be the one used as the Grignard reagent.

(a)

$$\underset{\displaystyle H_3C \quad\quad H}{\overset{\displaystyle O}{\overset{\displaystyle \|}{C}}} \quad \xrightarrow[\text{2. } H_2O/H^{\oplus}]{\text{1. ?}} \quad H_3C - \overset{OH}{\underset{CH_2CH_3}{\overset{|}{C}}} - H$$

(b)

$$\underset{\displaystyle H_3C \quad\quad CH_3}{\overset{\displaystyle O}{\overset{\displaystyle \|}{C}}} \quad \xrightarrow[\text{2. } H_2O/H^{\oplus}]{\text{1. ?}} \quad H_3C - \overset{OH}{\underset{CH-CH_3 \atop CH_3}{\overset{|}{C}}} - CH_3$$

(c)

$$\underset{\displaystyle CH_3CH_2 \quad\quad H}{\overset{\displaystyle O}{\overset{\displaystyle \|}{C}}} \quad \xrightarrow[\text{2. } H_2O/H^{\oplus}]{\text{1. ?}} \quad CH_3CH_2 - \overset{OH}{\underset{CH_3}{\overset{|}{C}}} - H$$

The alkyl groups in the squares are the groups which have been added. The corresponding Grignard reagents would therefore be:

(a)

$$CH_3CH_2 - MgI$$

(b)

$$IMg - \underset{CH_3}{\overset{|}{CH}} - CH_3$$

(c)

$$H_3C - MgI$$

8.4 Acid catalysed nucleophilic additions

It is found that some nucleophilic additions go faster if there is an acid catalyst present. Why should this be so? In order to answer that, we shall look at the reaction of propanone with hydrogen cyanide rather than with potassium cyanide. Hydrogen cyanide is a weak acid and so the following equilibrium takes place.

Hydrogen cyanide can therefore supply the necessary nucleophile for nucleophilic addition (cyanide ion) as well as a proton to catalyse the reaction. If the proton is catalysing the reaction, it may well be doing so by activating the aldehyde or ketone. Identify whether the proton is an electrophile or a nucleophile then identify which centre it would attack on propanone.

Answer

The proton is positively charged and is therefore an electrophile. An electrophile will react with a nucleophilic centre on propanone. The only nucleophilic centre is the oxygen.

We have identified a nucleophilic centre on propanone and we have an electrophilic proton. Show how a bond can form between these two centres and give the structure obtained.

Answer

One of the lone pairs on oxygen is used to form the bond to hydrogen. Since the two electrons are shared between hydrogen and oxygen, hydrogen effectively gains one electron, while oxygen effectively loses one. As a result, the proton loses its positive charge and the oxygen gains a positive charge.

What effect do you think a positively charged oxygen has on the neighbouring carbonyl carbon? Remember that oxygen is an electronegative atom.

Answer

Since oxygen is electronegative, it does not 'like' having a positive charge. It will therefore pull electrons to it even more strongly than before. This means that the polarity of the carbonyl bond is increased, making the carbonyl carbon even more electrophilic.

   We can illustrate this by invoking the following resonance. This results in the positive charge being shared between the oxygen and the carbon, making the carbon more electrophilic.

Increased
electrophilicity

If the carbonyl carbon is more electrophilic will this increase or decrease its reactivity to nucleophiles?

Answer

If the carbonyl carbon is more electrophilic, it means it is more electron-deficient. As a result, it will be more likely to react with something which is electron-rich—a nucleophile.

We have now seen how a proton can make a carbonyl group more reactive to nucleophiles. Give the full mechanism for the reaction of propanone with hydrogen cyanide.

Answer

Hydrogen cyanide is a weak acid and is partially dissociated to provide the proton and the cyanide nucleophile.

The carbonyl oxygen bonds to the proton using a lone pair of electrons. The resultant positive charge on oxygen means that oxygen pulls more strongly on the electrons making up the carbonyl double bond. This makes the carbonyl carbon more electrophilic and more reactive to cyanide. The cyanide uses its lone pair of electrons to bond to the carbonyl carbon. At the same time, the π electrons of the carbonyl bond move onto oxygen to restore the second lone pair and to neutralise the positive charge.

8.5 Reactivity of the carbonyl group—electronic and steric effects

It is found that aldehydes are more reactive to nucleophiles than ketones. For example, propanal will react faster than propanone with a cyanide ion.

Propanal   Propanone

There are two reasons for this—one electronic and one steric.

Electronic effects

The electronic factor should be familiar to you from previous sections in this text (see the discussion of the stability of carbocations in Section 5). However, let us take things one step at a time.

Since there is a difference in reactivity between aldehydes and ketones, it would make sense to identify any differences between these two functional groups. What differences can you identify?

Answer

Both groups have a carbonyl group (C=O) and so there is no difference there. The difference lies in what is attached to the carbonyl group. The ketone has alkyl groups attached to both sides of the carbonyl group while the aldehyde has only one alkyl group attached.

The ketone has two alkyl groups attached to the carbonyl carbon whereas the aldehyde has only one. For some reason, this has an effect on the reactivity of the carbonyl group. If this effect is electronic, it means that the alkyl groups are either electron-donating or electron-withdrawing. Which is correct? (You have already covered this under electrophilic additions.)

Answer

Alkyl groups are electron-donating.

If the alkyl groups are electron-donating, what effect do you think this has on the carbonyl carbon and its reactivity?

Answer

If alkyl groups are electron-donating, they will push electrons towards the carbonyl carbon. The carbonyl carbon is electron-deficient since the neighbouring oxygen has a greater share of the electrons in the double bond. However, alkyl groups attached to the carbonyl carbon push electrons towards this carbon which means that the carbonyl carbon is not so electron-deficient. This in turns means that it is less electrophilic and less reactive to nucleophiles.

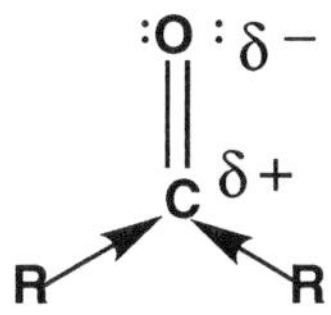

R = alkyl groups

Alkyl groups pushing
electrons towards
the carbonyl carbon

Explain now why an aldehyde (e.g. propanal) is more reactive than a ketone (e.g. propanone) to nucleophiles.

Propanal

Propanone

Answer

Propanal has only one alkyl group feeding electrons into the carbonyl carbon, whereas propanone has two. Therefore, the carbonyl carbon in propanal will be more electrophilic than the carbonyl carbon in propanone. The more electrophilic the carbon is, the more reactive it will be to nucleophiles. Therefore, propanal is more reactive than propanone.

Propanal   Propanone

Which of the following aldehydes would you expect to be more reactive to nucleophiles? Explain your answer.

Answer

The fluorinated aldehyde will be more reactive to nucleophiles.

The fluorine atoms are electronegative and have an inductive, electron-withdrawing effect on the neighbouring carbon, making it electron-deficient. This in turn leads to an inductive effect being felt by the neighbouring carbonyl carbon. Since electrons are being withdrawn, the electrophilicity of the carbonyl carbon is increased, making it more reactive to nucleophiles.

Trifluoromethyl group is electron-withdrawing and increases electrophilicity of carbonyl carbon.

Methyl group is electron-donating and decreases electrophilicity of carbonyl carbon.

Steric effects

Electronic factors are one reason why aldehydes are more reactive than ketones, but there are steric factors as well.

There are two ways of looking at how steric factors affect the reactivity of aldehydes and ketones. One is to look at the relative ease with which the attacking nucleophile can approach the carbonyl carbon. The other is to consider how steric factors influence the stability of the transition state leading to the final product.

Let us consider first the relative ease with which a nucleophile can approach the carbonyl carbon of an aldehyde and of a ketone. In order to do that, we must consider the bonding and the shape of the carbonyl functional group.

Identify the σ and π bonds in the following structures as well as any hybridised sp, sp^2 or sp^3 centres.

Bonding

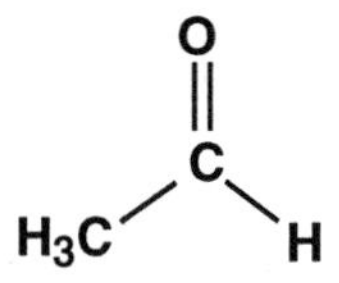

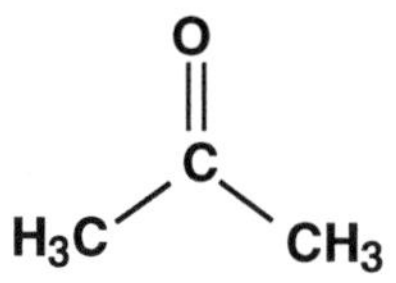

Hybridised centres

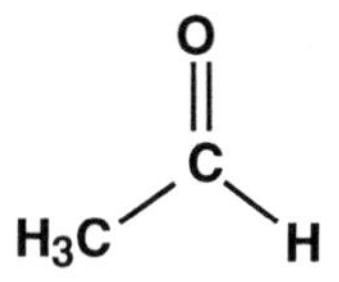

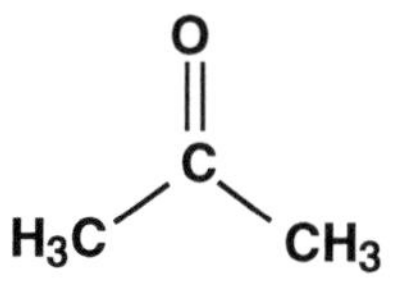

Answer

Bonding

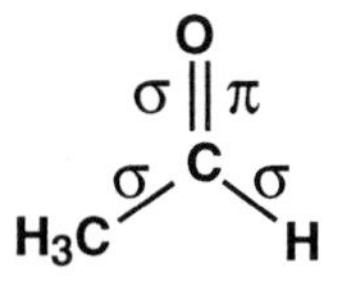

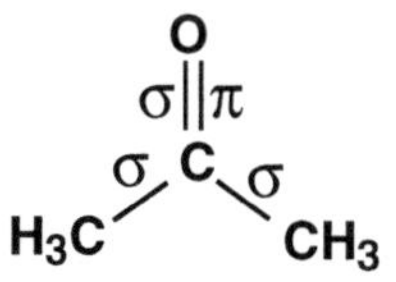

Hybridised centres

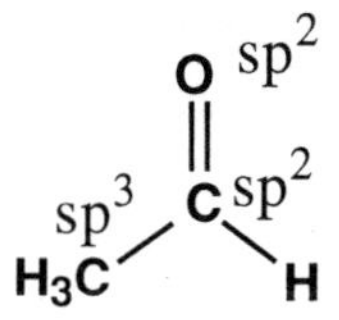

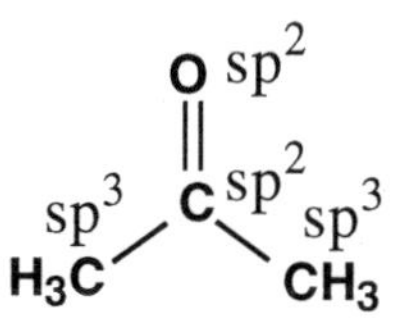

Describe and draw the shape of the above molecules.

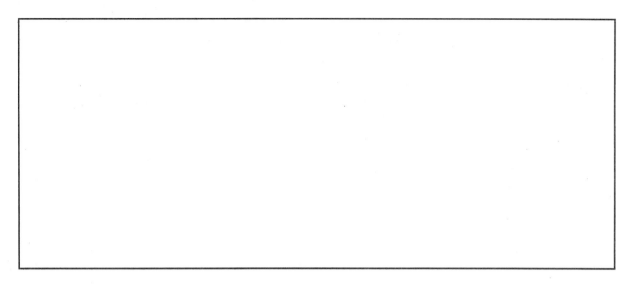

Answer

Both molecules have a planar carbonyl group. The atoms which are in the plane are circled below.

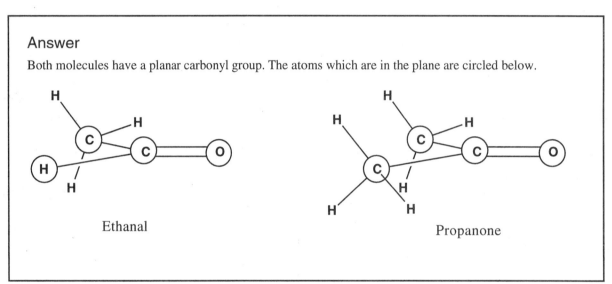

A nucleophile will approach the carbonyl group from above or below the plane. The diagram below shows a nucleophile attacking from above. Can you suggest any reason why it might be more difficult for the nucleophile to attack the carbonyl group in propanone than the carbonyl group in ethanal?

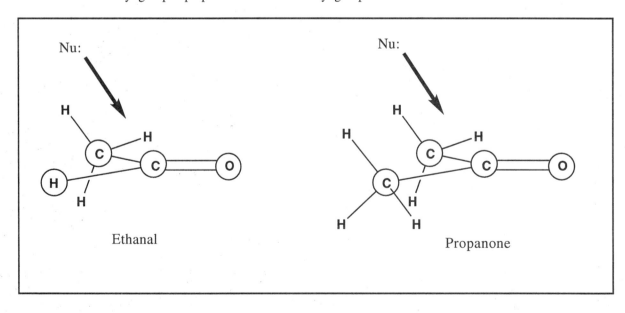

Answer

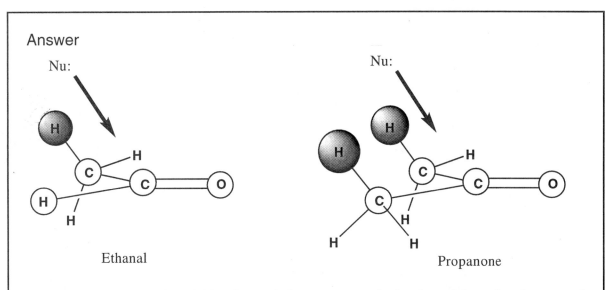

Ethanal

Propanone

The hydrogen atoms on the neighbouring methyl groups are not in the plane of the carbonyl group and could hinder the approach of a nucleophile. This is more significant for propanone where there are methyl groups on both sides of the carbonyl group.

We shall now look at how steric factors affect the stability of the transition state leading to the final product.

In order to understand that, we have to look at what happens to the size and shape of the aldehyde or ketone when it undergoes nucleophilic addition.

Compare the shape of propanone with the cyanohydrin product resulting from nucleophilic addition of HCN. Draw both structures.

If you have problems visualising the 3D shapes of molecules, it is worthwhile buying molecular models. These will become essential as you continue your studies in Chemistry.

Answer

Propanone is a flat molecule whereas the cyanohydrin molecule is tetrahedral.

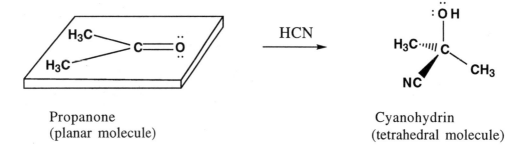

Propanone
(planar molecule)

Cyanohydrin
(tetrahedral molecule)

(Note how a tetrahedral molecule is represented on the written page. Bonds in the plane of the page are written normally. Bonds coming out the page towards you are drawn as a solid wedge. Bonds going into the page are drawn as a hatched wedge.)

There is a marked difference in shape between the starting ketone and the cyanohydrin product. There is also a difference in the space available to the groups attached to the reactive site—the carbonyl carbon. Compare the two structures and decide whether one is more crowded than the other.

Answer

Again models help here, but nevertheless you should be able to see that the tetrahedral molecule is more crowded than the planar molecule. In the tetrahedral product, there are four groups crowded round a central carbon, whereas in the planar starting material, there are only three groups attached to the carbonyl carbon, and they are quite far apart.

Compare now the reaction of propanal and propanone with HCN and the products which would be obtained. Which of these would be more crowded?

Answer

The cyanohydrin product from propanal will be less crowded than the cyanohydrin product from propanone since one of the groups round the central carbon is a small hydrogen atom.

The ease with which nucleophilic addition takes place will depend on the ease with which the transition state is formed. In nucleophilic addition, the transition state is thought to resemble the tetrahedral product more than the planar starting material. Which of the above starting materials do you think would be more reactive and why?

Answer

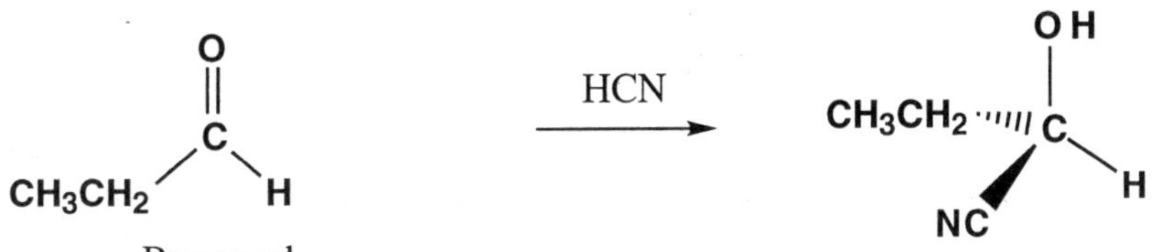

Propanal

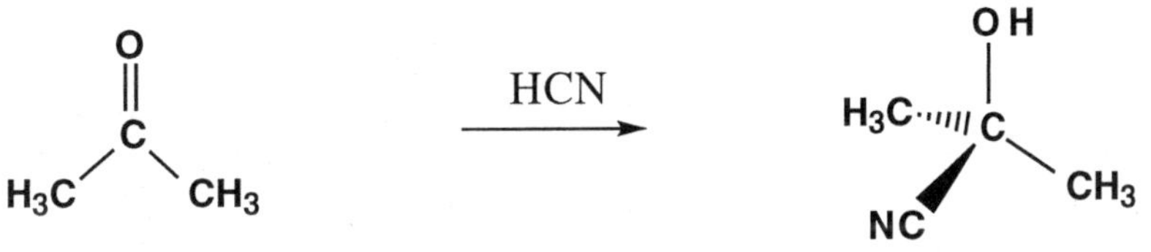

Propanone

Propanal should be more reactive since the transition state resulting from nucleophilic addition will be less crowded. The more crowding there is, the less stable the transition state.

The bigger the alkyl groups, the bigger the steric effect. Which of the following ketones would you expect to be more reactive to potassium cyanide?

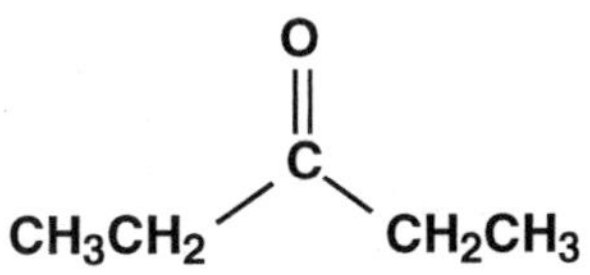

3-Pentanone

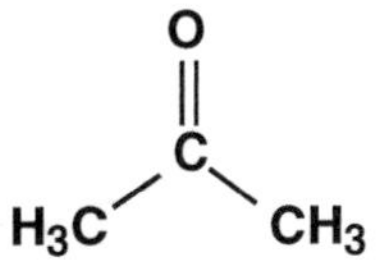

Propanone

Answer

Propanone should be more reactive. The ethyl groups in 3-pentanone are larger than the methyl groups in propanone. They would hinder the approach of the nucleophile to a greater extent and also increase the crowding in the transition state.

8.6 Examination technique

The following question is the sort you might expect to get in an exam. Answer all parts before looking at the answers.

(i) Give the correct names for the following structures.

(a) 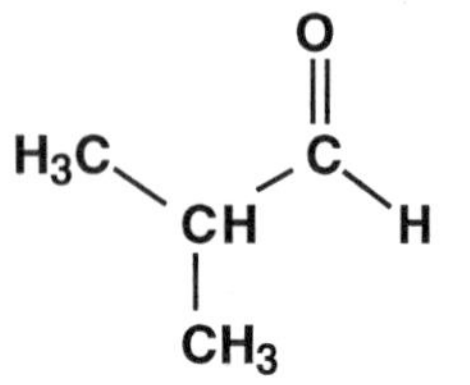   (b) 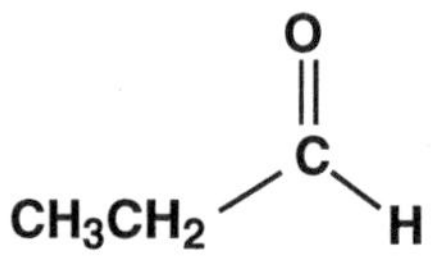   (c) 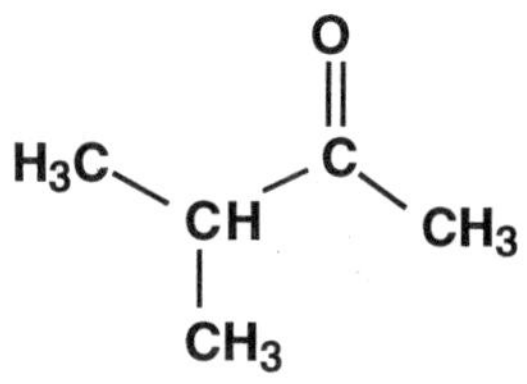   (3)

(a)   (b)   (c)

(ii) Identify the number of non equivalent protons and carbons you would observe for each of the above structures.   (3)

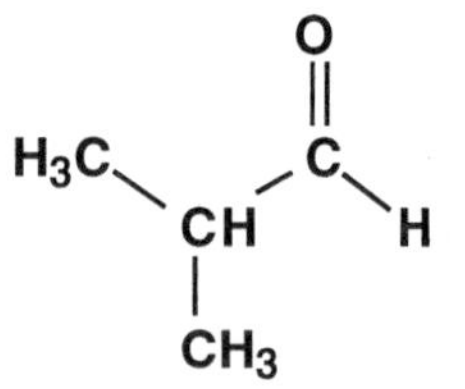

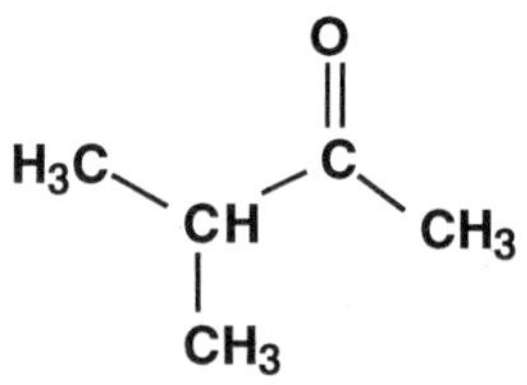

No. of non-
equivalent
hydrogens

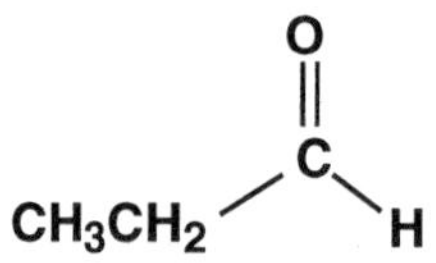

No. of non-
equivalent
carbons

(iii) Draw the structure of the product obtained if compound (b) was treated with methyllithium (CH_3Li), then water. (Methyllithium is similar to methylmagnesium iodide in that it provides the equivalent of a methyl carbanion.)   (3)

(iv) Give the mechanism for this reaction and explain the steps involved. (10)

(v) Explain whether the above compounds (a) and (c) would react at the same rate with methyllithium or not?

(6)

There is quite a technique to answering exam questions, so we shall look at this exam question in some detail with respect to the answer and the way it might be marked by an examiner.

Answer

It is important to try and 'read the examiners mind' when answering exam questions. Look carefully at the wording of the questions and the marks given. These should provide you with a clue to what the examiner is looking for and (very importantly) in what detail. Too often, students misread questions or take no notice of the marks awarded. If a question has only 1 mark, then it is a waste of time writing a full page on that question, when clearly the examiner is only wanting a short answer. On the other hand, a question demanding several marks requires more than a one line answer. Let us now look at the answers for part (i).

(i)

(a) 2-Methylpropanal, (b) Propanal, (c) 3-Methylbutanone.

This question is straightforward. There are 3 marks awarded, so clearly there is 1 mark for each correct name. Give yourself 1 mark for each correct answer. Note that there is no need to identify the position of the functional group in any of these molecules since they cannot be anywhere else. An aldehyde functional group will always be at position 1. The ketone group of butanone can only be at position 2. It is unlikely that you would lose marks for putting something like 3-methyl-2-butanone, but you would certainly lose marks if you put 2-methyl-3-butanone. If you made a mess of this, go back to the section on nomenclature in '*Beginning Organic Chemistry 1*' and go through that again.

(ii)

There are 3 marks for part (ii) so there is 1 mark for each structure. For each structure you would therefore get half a mark if you get the number of non-equivalent hydrogens correct and another half mark if you get the number of non-equivalent carbons right.

Structure	(a)	(b)	(c)
No. of non-equivalent hydrogens	3	3	3
No. of non-equivalent carbons	3	3	4

This is all you have to put down to answer the question. However, if the question had asked you to 'identify the non-equivalent protons and carbons for each structure', then you would have to put down the structures showing which protons and carbons are non-equivalent. For example, the non-equivalent carbons for structure (a) would be as shown below:

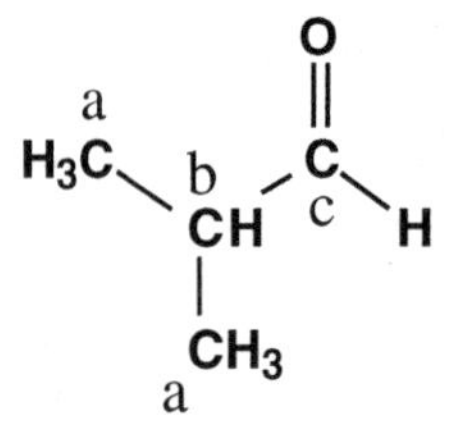

three non-equivalent carbons

(iii)

This question is very specific in what it wants. There are only 3 marks, so a lengthy explanation is not required. Since there are 3 marks, the examiner is presumably wanting the intermediate in the reaction as well as the final structure.

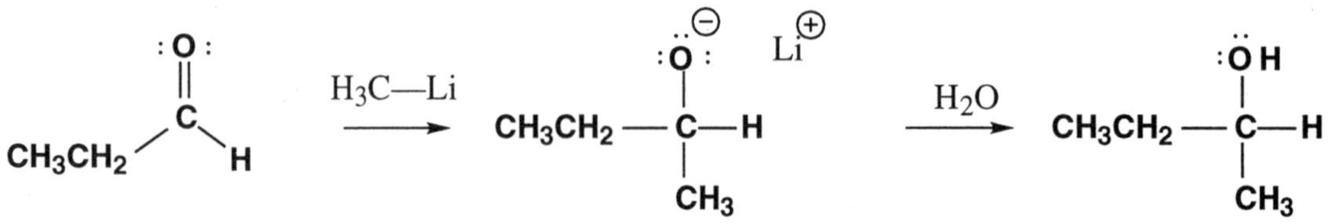

cont.

(iv)

This part has a total of 10 marks and is therefore the 'meat' of the question. It is clearly worth spending a bit of time on this part. The question asks for the mechanism and an explanation, so you would give the complete mechanism, then add a full explanation, as follows.

Mechanism

π bond
beaking

Explanation

'The carbonyl bond of the aldehyde is polarised because of the electronegative nature of the oxygen. Since oxygen is more electronegative than carbon, the electrons in the carbonyl bond spend more of their time nearer oxygen than carbon. As a result, the oxygen is slightly negative and the carbon is slightly positive. The carbon is therefore electron-deficient and electrophilic.

Electrophilic centres will react with nucleophiles. The nucleophile in this case is the methyl carbanion. The carbon of the methyl carbanion has a lone pair of electrons and is negatively charged, so this is the nucleophilic centre. The lone pair of electrons on the methyl carbanion is used to form a bond to the electrophilic carbonyl carbon. As this new bond forms, one of the bonds already present to the carbonyl carbon must break since carbon is only allowed a maximum of four bonds. The weakest bond will break and this will be the π bond. Both electrons in the π bond will move onto oxygen to give it a third lone pair of electrons and a negative charge.

On adding water, this negatively charged oxygen can act as a nucleophile and use a lone pair of electrons to form a bond to one of the protons on water. As it does so, the bond from hydrogen to oxygen on water is broken since hydrogen can only have one bond. Both electrons in the bond end up on the oxygen to give a third lone pair and a negative charge.'

You may wonder how the 10 marks are awarded in a question like this. For example, it is not clear whether the mechanism and the explanation will be given 5 marks each or whether there will be a different weighting. The way that the question is phrased gives no clues and so it is best to assume that the weighting of the marks will be approximately equal. Often examiners are reluctant to be too precise with marks since they wish to be flexible. For example, a student may give a superb explanation of the mechanism, but perhaps have a couple of trivial errors in the mechanism itself. By having some flexibility, the examiner can take account of the good explanation and ignore the errors in the mechanism. It is difficult to say exactly how a question of this sort will be marked since each examiner may be looking for different key phrases or key points.

The following is an example of how one examiner marked the above answer. Each star represents a key point.

cont.

π bond
beaking

This particular examiner has decided to give credit for the following points. Are the curly arrows drawn from the right place and are they pointing to the right place? Have the correct charges been put in? Have any new bonds or lone pairs been correctly put in? By asking these three questions, the examiner has been looking for 13 key points in the answer. Let us now look at the explanation and how it was marked. The key phrases which the examiner was looking for are underlined.

Explanation

'The carbonyl bond of the aldehyde is polarised because of the electronegative nature of the oxygen. Since oxygen is more electronegative than carbon, the electrons in the carbonyl bond spend more of their time nearer oxygen than carbon. As a result, the oxygen is slightly negative and the carbon is slightly positive. The carbon is therefore electron-deficient and electrophilic.

Electrophilic centres will react with nucleophiles. The nucleophile in this case is the methyl carbanion. The carbon of the methyl carbanion has a lone pair of electrons and is negatively charged so this is the nucleophilic centre. The lone pair of electrons on the methyl carbanion is used to form a bond to the electrophilic carbonyl carbon. As this new bond forms. one of the bonds already present to the carbonyl carbon must break since carbon is only allowed a maximum of four bonds. The weakest bond will break and this will be the π bond. Both electrons in the π bond will move onto oxygen to give it a third lone pair of electrons and a negative charge.

On adding water, this negatively charged oxygen can act as a nucleophile and use a lone pair of electrons to form a bond to one of the protons on water. As it does so, the bond from hydrogen to oxygen is broken since hydrogen can only have one bond. Both electrons in the bond end up on the oxygen to give a third lone pair and a negative charge.'

The examiner has decided to look for 11 key points, words or phrases in the answer and to give credit for these. You may wonder why other parts of the explanation have not been credited. This is because the examiner has decided that credit for those points was already given in the mechanism. However, it is still important to give the full explanation. You cannot predict what the examiner will be looking for in both parts. Also, you may have forgotten to put an arrow into the mechanism. The examiner may decide to ignore that omission if your explanation shows that you understand what is going on.

Overall, this particular examiner identified 13 points in the mechanism and 11 points in the explanation. This clearly does not come to 10, so he/she would probably give half a mark for each point. This still does not come to 10 so the examiner might make a correction or decide to give full marks if the student got 20 out of the 24 points looked for. That is one way by which an examiner might mark, where the examiner is 'marking by a formula'. Another method of marking is where the examiner marks by 'general impression', i.e. he or she reads through your answer and then marks you out of 10 based on how clearly you argue your case. However, this method of marking is becoming less favoured these days.

(v)

This part has 6 marks attached to it, so a full explanation is required. The following answer would get full marks.

'The three compounds would not react at the same rate. Compound (b) would react faster than compound (a). Compound (a) would react faster than compound (c).'
'The rate of reaction in nucleophilic addition is affected by electronic and steric factors.'

cont.

'Alkyl groups attached to the carbonyl carbon have an electronic effect since they push electrons towards that centre and make it less electrophilic and therefore less reactive. Therefore, compound (c) would be the least reactive compound since it has two alkyl groups feeding electrons to the carbonyl centre. Compounds (a) and (b) have one alkyl group feeding electrons to the carbonyl centre.'

'There is also a steric factor. The product of the reaction is tetrahedral with four groups round a central carbon, compared to the starting material which is planar with three groups round the carbonyl carbon. The product is therefore more crowded. The transition state is thought to resemble the product more than the starting material and so alkyl groups which are likely to increase crowding in the transition state will hinder the reaction.'

Planar   Tetrahedral

'Compound (c) has two alkyl groups attached to the carbonyl group so the transition state from (c) will be more crowded than the transition states from (a) or (b). Therefore, compound (c) is least reactive. The alkyl group on (a) is larger than the one on (b). Therefore structure (a) would react more slowly than (b) since its transition state is more crowded.'

'Larger alkyl groups will also hinder the approach of the nucleophile towards the reaction centre.'

Summary

- Aldehydes and ketones undergo nucleophilic addition reactions with nucleophiles.

- The carbonyl group of an aldehyde or ketone group is polarised such that the carbonyl carbon atom is electrophilic.

- Nucleophiles form a new bond to the electrophilic carbonyl carbon with simultaneous breaking of the carbonyl π bond.

- The product from nucleophilic addition is a more crowded tetrahedral structure compared to the planar carbonyl group.

- The transition state for nucleophilic addition is thought to resemble the product more than the starting material.

- The rate of nucleophilic addition is affected by electronic and steric factors.

- Electron-donating alkyl groups lower the rate of reaction.

- Electron-withdrawing substituents increase the rate of reaction.

- Large bulky substituents lower the rate of reaction.

- Ketones are less reactive than aldehydes to nucleophilic addition.

SECTION 9

Nucleophilic Substitutions of Carboxylic Acid Derivatives

Nucleophilic substitutions of carboxylic acid derivatives

9.1 Nucleophilic substitutions—introduction

Nucleophilic substitutions are reactions which take place with alkyl halides or with carboxylic acid derivatives (acid chlorides, anhydrides, esters and amides). The following reactions are examples of nucleophilic substitutions.

$$I - CH_3 \quad + \quad HO^{\ominus} \quad \longrightarrow \quad HO - CH_3 \quad + \quad I^{\ominus}$$

Alkyl halide   **Alcohol**

Acid chloride $+$ $HO^{\ominus}$ $\longrightarrow$ Carboxylic acid $+$ $Cl^{\ominus}$

Ester $+$ $HO^{\ominus}$ $\longrightarrow$ Carboxylic acid $+$ $^{\ominus}OCH_2CH_3$

Acid anhydride $+$ $HO^{\ominus}$ $\longrightarrow$ Carboxylic acid $+$

Amide $+$ $HO^{\ominus}$ $\longrightarrow$ Carboxylic acid $+$ $^{\ominus}NHCH_3$

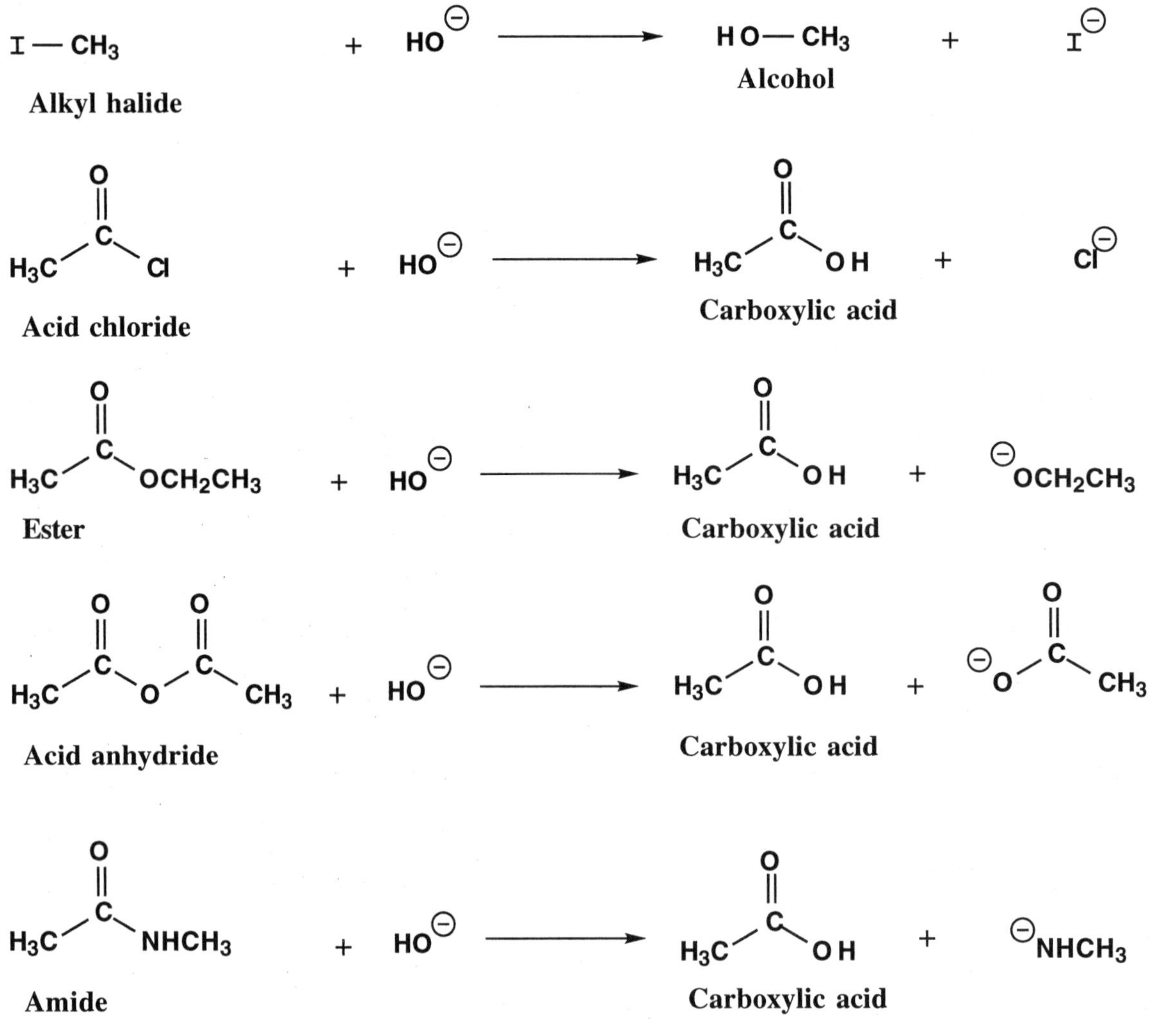

Look at the reactions given above and suggest why they are all called nucleophilic substitutions.

Answer

The reactions above are called nucleophilic substitutions because one nucleophile (the hydroxide ion) displaces or substitutes another nucleophile from the molecule.

Although all these reactions are nucleophilic substitutions, the mechanisms involving alkyl halides and carboxylic acid derivatives are different. In this section we shall look at what happens when carboxylic acid derivatives react with nucleophiles. In Section 10 we shall look at the nucleophilic substitution of alkyl halides.

9.2 Nucleophilic substitution of acid chlorides with the methoxide ion

The four carboxylic acid derivatives all take part in nucleophilic substitutions by the same mechanism. We shall concentrate on acid chlorides, but the same principles hold for acid anhydrides, esters and amides. Ethanoyl chloride (also called acetyl chloride) is an acid chloride, with the formula shown below.

Identify any nucleophilic or electron-rich centres in the molecule.

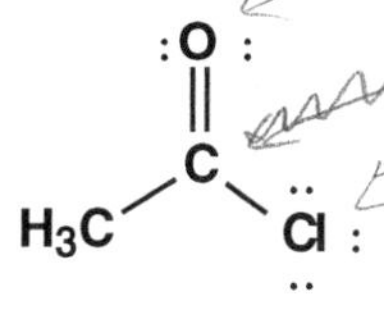

Answer

Nucleophilic centres are electron-rich atoms. Electron-rich atoms either have a negative charge or have lone pairs of electrons. Therefore, the oxygen and the chlorine are nucleophilic centres.

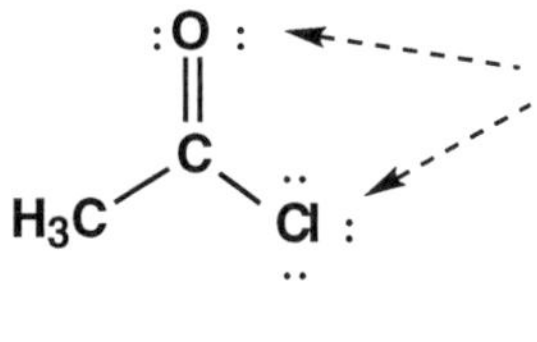

Both oxygen and chlorine are more electronegative than carbon. What effect has this on the polarity of the molecule? Identify any electrophilic centres as a result of this polarity.

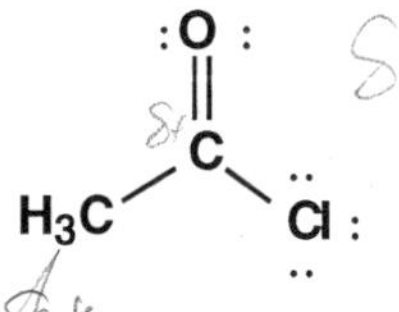

Answer

Since oxygen and chlorine are more electronegative than carbon, the electrons in the carbon–oxygen bond and the carbon–chlorine bond will be pulled towards the more electronegative atoms, resulting in the oxygen and chlorine gaining a partial negative charge. The carbon will of course gain a partial positive charge. Since the carbon is partially positive, it is electron-deficient and is therefore an electrophilic centre.

Electrophilic
centre

Would you expect the carbonyl carbon of an acid chloride to be more or less electrophilic than the carbonyl carbon of a ketone? Give an explanation for your answer.

Answer

The carbonyl carbon of an acid chloride should be more electrophilic since there are two electronegative atoms withdrawing electrons from it compared with only one for the ketone.

Taking this into account, would you expect acid chlorides to be more or less reactive than ketones?

Answer

Acid chlorides are more reactive to nucleophiles since the carbonyl carbon is more electronegative and 'hungrier' for electron-rich molecules (nucleophiles).

We have identified the electrophilic centre in ethanoyl chloride and we know that this is the centre that will react with a nucleophile in the formation of a new bond. You should know from previous sections that the nucleophile will provide the electrons for this new bond. Let us look at what happens if ethanoyl chloride is treated with the methoxide ion.

Identify the nucleophilic centre in the methoxide ion and the electrons which will be used to form the new bond.

Answer

The oxygen of the methoxide ion is negatively charged and is electron-rich. Therefore, the oxygen is the nucleophilic centre. One of the lone pairs on oxygen will be used for the new bond.

If the methoxide ion forms a new bond to the carbonyl carbon on the acid chloride, how many bonds would that carbon have? What conclusion do you come to?

Answer

The carbonyl carbon has four bonds already, so another bond to it would make five. However, carbon cannot have more than four bonds. Therefore, one of the original four bonds will break, as the new one is formed.

If one of the four original bonds breaks to allow the new bond, which bond will break? Give reasons for your answer and explain where the electrons in that bond will go.

Answer

The weakest bond will break, and this is the π bond of the carbonyl group. This bond is weaker than the other three σ bonds, since it is formed from the weaker side-on overlap of p orbitals. The two π electrons will go onto the oxygen as a third lone pair, thus giving oxygen a negative charge.

We have argued logically what will happen if a nucleophile such as the methoxide ion attacks ethanoyl chloride. Summarise everything with a mechanism using curly arrows to show the movement of the electrons. Show the structure of the intermediate formed.

Answer

Intermediate

If you look back to the reaction of ketones and aldehydes with nucleophiles, this is exactly the same first step as that involved in nucleophilic addition. However, once the intermediate is formed something different happens. We know that the overall reaction is a nucleophilic substitution rather than a nucleophilic addition. We also know that it is the chloride ion which is substituted. We now have to work out how that chloride ion is lost. The chloride anion has four lone pairs of electrons and a negative charge. It only has three lone pairs of electrons in the intermediate so where do you think the two electrons for the extra lone pair come from?

Answer

The most likely source for the fourth lone pair is the bond between the carbon and the chlorine. We know that this bond has to break, and so both electrons in the bond must move onto chlorine.

Draw a mechanism to represent this and show the structures obtained.

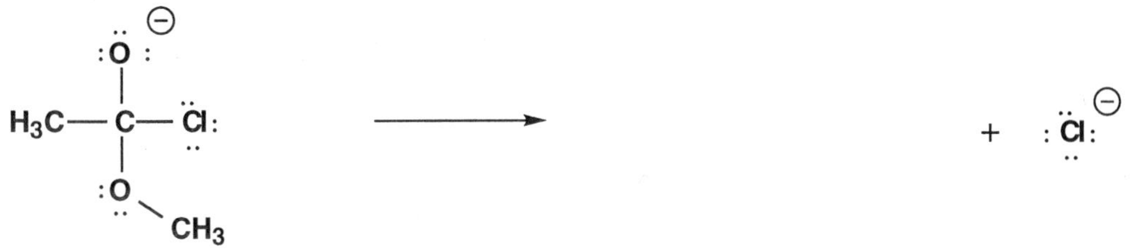

Answer

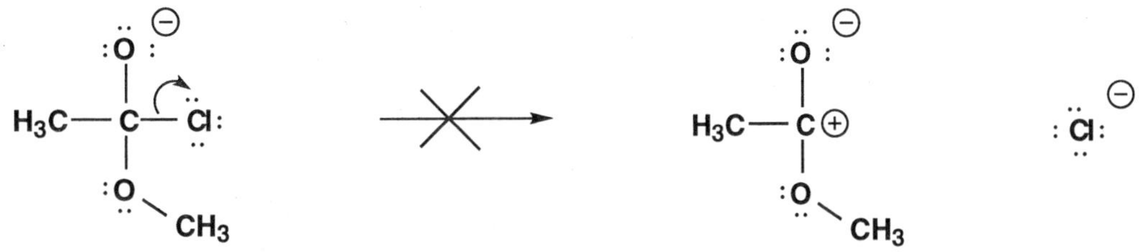

Unstable resonance
structure

Based on what we have argued so far, the above answer is probably what you would have obtained. It certainly explains how we get the chloride ion leaving the molecule, but the mechanism is fatally flawed. Why?

Answer

The mechanism above would mean that we are creating a positive charge on the carbon. This implies that we are trying to remove a negatively charged ion from a positive centre which is very difficult. Furthermore, the positively charged carbon has two electronegative oxygens attached to it which would destabilise a positive charge and make it highly unstable.

How then could you draw a mechanism which would allow the C–Cl bond to break but which would not cause the carbon to gain a positive charge?

Let us look at the structure of the intermediate more closely. Identify the nucleophilic and electrophilic centres of this molecule and suggest whether you think this is a stable arrangement of atoms.

Answer

There are three electron-rich (nucleophilic) atoms attached to the same carbon. Since all three of these carbons are electronegative, it means that this carbon is very electrophilic indeed.

With such a strong electrophilic centre, this intermediate will be unstable and undergo further reaction.

Since the electrophilic carbon is desperately short of electrons, it needs to gain electrons from somewhere. Where is the most readily available source of electrons? What do you think is likely to happen next?

Answer

The negatively charged oxygen is the most readily available source of electrons. A lone pair of electrons from that oxygen can come in to reform the π bond between the carbon and oxygen.

If the π bond reforms, how many bonds will carbon have? Is this possible? If not, what must happen?

Answer

If the π bond reformed and nothing else happened, the central carbon would end up with five bonds. This is not possible, so one of the four bonds already present must break.

Which of the four bonds in the above intermediate is the most likely to break?

Answer

All four bonds are σ bonds. We can discount the carbon–carbon bond since this is a strong bond. It is also impossible for the σ bond of the carbonyl bond to break if the π bond is being created, so that leaves the C–OCH$_3$ bond and the C—Cl bond. Both of these bonds are weaker than carbon–carbon or carbon–hydrogen bonds. Therefore, both are likely to break. However, if the C–OCH$_3$ bond broke, we would end up with what we started with. If the C–Cl bond broke, we would obtain the correct product.

Draw a mechanism showing where the electrons all end up, assuming that the C–Cl bond is broken.

Answer

The mechanism is as follows.

We now have the observed products. However, we cannot just say that the C–Cl bond breaks because that would give us the correct product. We have to come up with an explanation which would explain why the C–Cl bond is the more likely bond to break.

The best explanation for this involves the leaving groups which would be formed from either process. Identify the leaving groups if the C–Cl bond or the C–OCH$_3$ bond is broken.

Answer

If the C–Cl bond breaks, the leaving group is the chloride ion.
If the C–OCH$_3$ bond breaks, the leaving group is the methoxide ion.

Chloride   Methoxide

Clearly, the chloride ion must be the better leaving group since this gives us the observed products. Why is it a better leaving group than the methoxide ion? Would you expect the better leaving group to be more stable or less stable than the poorer leaving group?

Answer

We would expect the better leaving group to be more stable since it would be more easily formed.

The chloride ion must be more stable than the methoxide ion. Why should this be so?

Answer

Chlorine is more electronegative than oxygen and can stabilise the negative charge much better.

This is also reflected in the relative basicities of the chloride ion and the methoxide ion. The chloride ion is a very weak base and is fully ionised in water, whereas the methoxide ion is a strong base and would react with water to give methanol plus the hydroxide ion.

We have now gone through the various stages of the mechanism. See if you can draw the complete mechanism below.

Answer

We now have mechanisms to show how nucleophilic addition takes place on a ketone or aldehyde and how nucleophilic substitution takes place on an acid chloride. But why the difference? We have seen above what happens when the methoxide ion is treated with ethanoyl chloride. Remembering what you have learnt about ketones and aldehydes, what product would you expect if propanone is treated with the methoxide ion, then water. What is the mechanism?

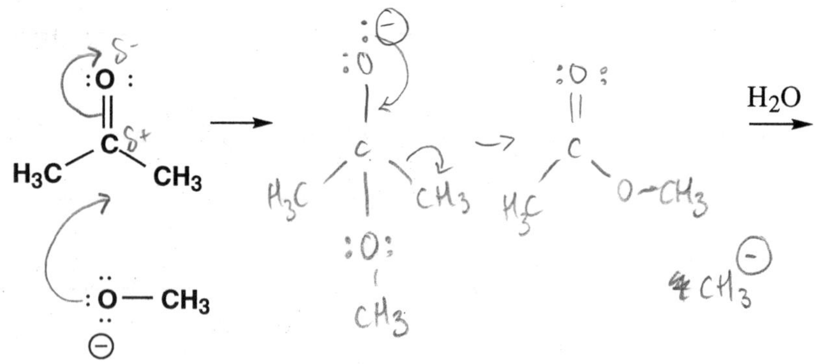

Answer

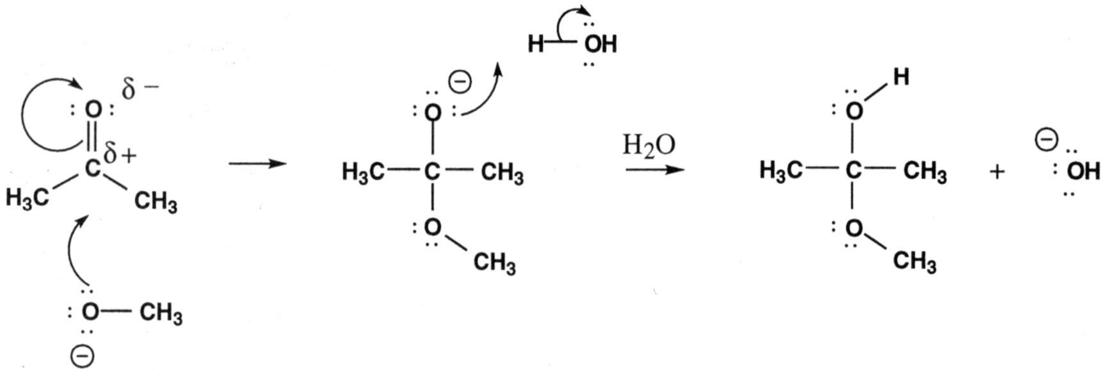

Draw the mechanism which would be involved if nucleophilic substitution took place between propanone and a methoxide ion. Explain why this reaction is impossible.

Answer

Nucleophilic substitution of a ketone is not possible since it would require the breaking of a strong carbon–carbon bond. Furthermore, the methyl carbanion is a highly reactive species and a very strong base. It is not easily formed and is a very poor leaving group.

Give the hypothetical mechanism for the nucleophilic substitution of ethanal (acetaldehyde) with the methoxide ion. Explain why this reaction mechanism is not possible.

Answer

Nucleophilic substitution of an aldehyde is not possible since it would require the breaking of a strong carbon–hydrogen bond. Furthermore, the hydride anion is a highly reactive species and a very strong base. It is a very poor leaving group and not easily formed.

9.3 Reaction of acid chlorides with nucleophiles other then methoxide

We have looked at the nucleophilic substitution of an acid chloride with the methoxide ion. Acid chlorides are reactive and will undergo reaction with a variety of nucleophilic anions. These reactions are especially useful in synthesising other types of carboxylic acid derivatives. Give the reaction products which you would expect from the following reactions and state what type of functional group is formed.

(a) $H_3C-C(=O)-Cl$ NaOEt   Sodium ethoxide $\longrightarrow$

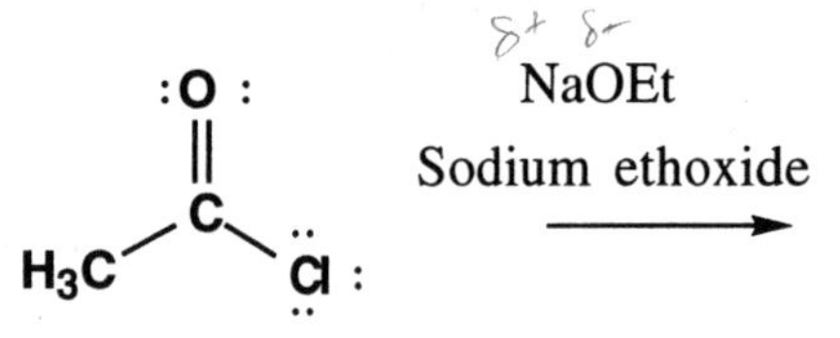

(b) $H_3C-C(=O)-Cl$ $CH_3CO_2^{-} Na^{+}$ Sodium ethanoate $\longrightarrow$

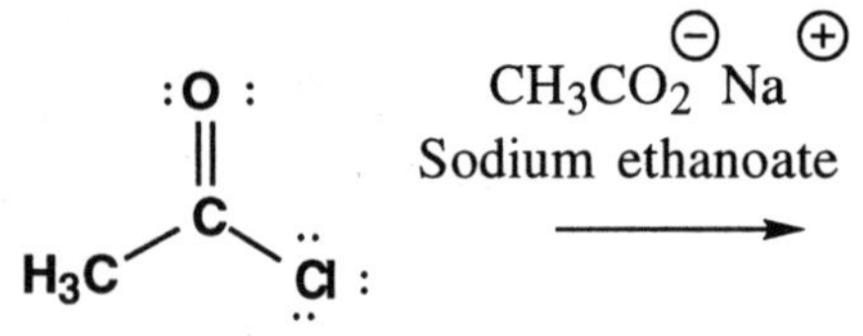

(c) $H_3C-C(=O)-Cl$ $CH_3CH_2-NH^{-} Na^{+}$ $\longrightarrow$

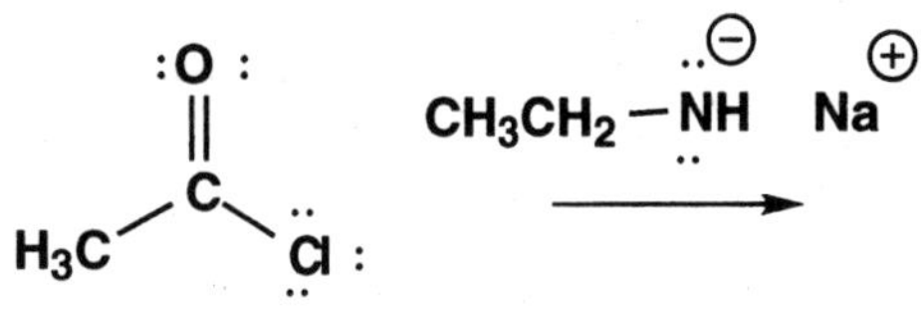

Answer

(a) $H_3C-C(=O)-Cl$ $+$ $[CH_3CH_2-O^{-}]$ Na^{+} $\longrightarrow$ $H_3C-C(=O)-[OCH_2CH_3]$ $+$ Na^{+} $:Cl:^{-}$

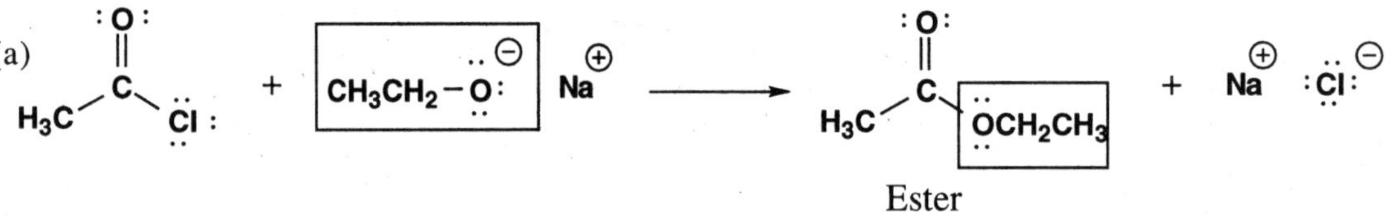

Ester

(b) $H_3C-C(=O)-Cl$ $+$ $[H_3C-C(=O)-O^{-}]$ Na^{+} $\longrightarrow$ $H_3C-C(=O)-[O-C(=O)-CH_3]$ $+$ Na^{+} $:Cl:^{-}$

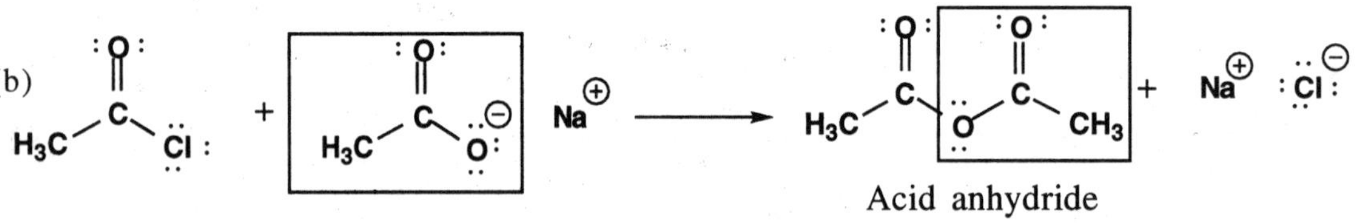

Acid anhydride

(c) $H_3C-C(=O)-Cl$ $+$ $[CH_3CH_2-NH^{-}]$ Na^{+} $\longrightarrow$ $H_3C-C(=O)-[NHCH_2CH_3]$ $+$ Na^{+} $:Cl:^{-}$

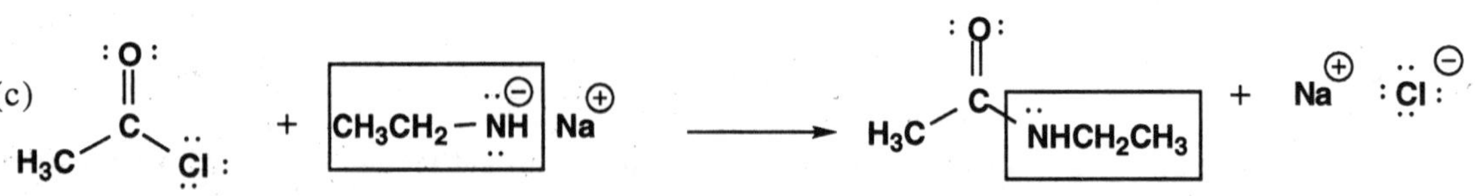

Amide

9.4 Reaction of acid chlorides with alcohols by nucleophilic substitution

Acid chlorides are reactive enough to react with uncharged weaker nucleophiles such as alcohols and amines. For example, ethanoyl chloride will react with methanol to give the same ester obtained from the reaction of ethanoyl chloride with sodium methoxide.

Ester

Propose a mechanism for this reaction. Remember to identify the nucleophilic centre of the nucleophile and to identify the electrons which will be used for the new bond.

Answer

The nucleophilic centre of methanol is oxygen since it is electron-rich and has two lone pairs of electrons. One of these lone pairs of electrons is used to form the bond to the electrophilic carbon centre of the acid chloride. As this new bond forms, the weak π bond in the carbonyl bond breaks with both electrons moving onto the carbonyl oxygen, resulting in a negative charge on the oxygen.

Note that the methanol oxygen gains a positive charge since it has effectively lost an electron.

The positively charged oxygen is not very stable and so the next stage in the mechanism is the loss of the proton on oxygen. Both electrons in the O–H bond move onto the oxygen to restore a second lone pair of electrons and to neutralise the charge. The methanol is usually present as solvent as well as reactant and can aid the process by acting as a nucleophile to 'pick off' the proton.

The final stage in the mechanism is the same as before. One of the lone pairs on the negatively charged oxygen comes in to reform the π bond and the C=O group. As this π bond is formed, the C–Cl bond breaks. Both electrons move to the chlorine to form a fourth lone pair of electrons.

Finally, the chloride anion can remove a proton from $CH_3OH_2^+$ to form HCl.

The above mechanism appears very complicated but it is essentially the same mechanism involved in the reaction of ethanoyl chloride with sodium methoxide. The only difference is the fact that we have a proton to 'get rid of' at some time during the reaction mechanism.

The overall reaction mechanism can be summarised as below.

You might well ask why the steps are written in this order? For example, why is the proton not lost after reforming the π bond rather than before it (see below)?

There is a good reason for favouring the first mechanism over the second one. Can you suggest what it might be? (Hint: look at the second step.)

Answer

The problem with the second mechanism is the second step. If the proton is still attached to the oxygen, then this atom is positively charged and makes an excellent leaving group. Therefore, when the carbonyl group is reformed, it is more likely that methanol will be thrown out rather than the chloride ion. This would give back the starting materials and does not explain how we get the observed products.

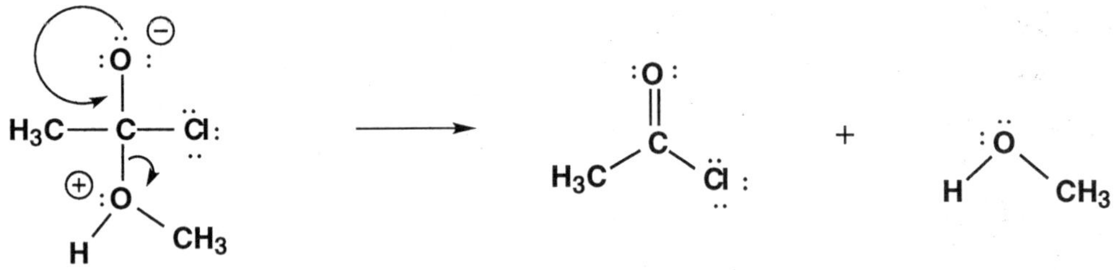

The first mechanism is therefore better since the proton and the positive charge are lost before the π bond reforms.

9.5 Reaction of acid chlorides with amines by nucleophilic substitution

Acid chlorides react with ammonia and amines to give amides. Give the products you would expect from the following reactions.

(a)

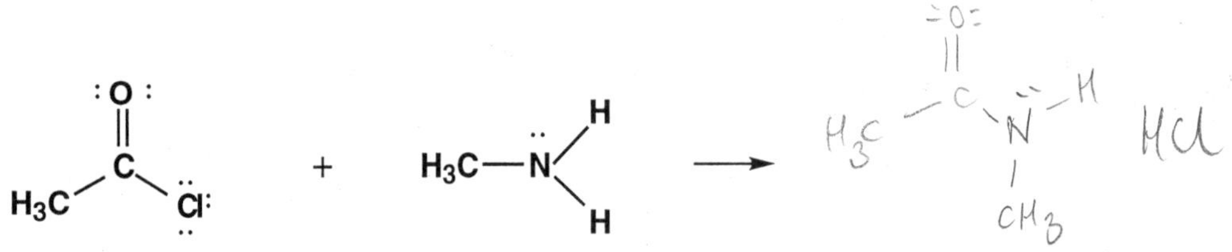

(b)

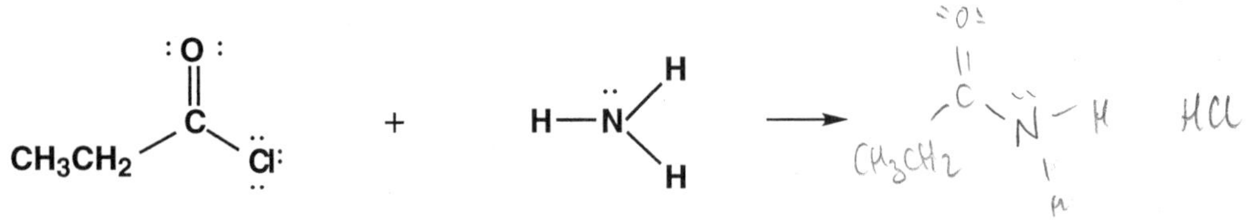

(c)

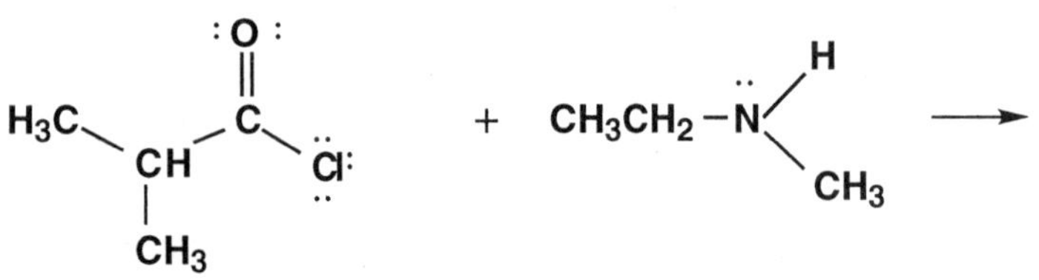

Answer

(a)

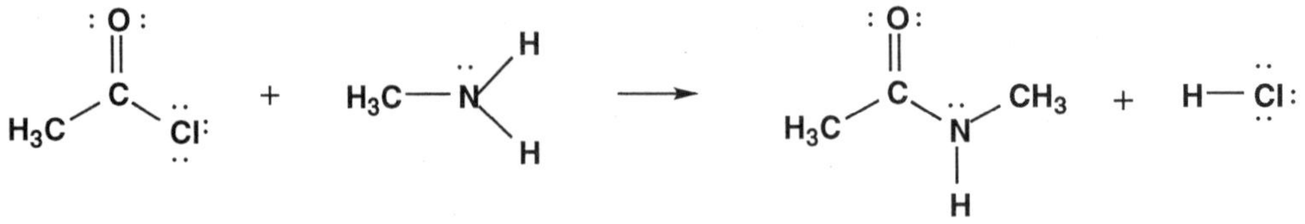

cont.

If you found the last problem slightly tricky, you might be getting confused by the non-essential parts of the molecule. If you are faced with any question asking you to identify the product of a reaction, then follow this procedure.

(1) Identify the functional groups which are reacting in both molecules.

(2) Simplify the remainder of the molecule by using the shorthand (R) for various alkyl substituents.

(3) Identify the reaction which is taking place and the products which are formed.

(4) Replace the R groups with the actual substituents.

As an example consider the last problem above.

• Stage 1. Identify the reacting functional groups.

• Stage 2. Simplify the molecules.

Let R = CH₃ — CH   R' = CH₃   R'' = CH₂CH₃
                 |
               CH₃

Therefore:

- Stage 3.   Identify the reaction and reaction products.

- Stage 4.   Replace R groups.

(c)

9.6 Nucleophilic substitutions of acid anhydrides, esters and amides

We have looked at nucleophilic substitutions of acid chlorides and how these reactions can give other carboxylic acid derivatives. These other derivatives can also undergo nucleophilic substitution but are less reactive than acid chlorides. The order of reactivity is as follows.

Acid chloride   **Acid anhydride**   **Ester**   **Amide** R"
(most reactive)   **(least reactive)**

Decreasing reactivity

We are going to investigate why there is a difference in reactivity.
The general equation for the reaction of an acid chloride with a charged nucleophile is as follows.

Give the general equations for the nucleophilic substitutions of a charged nucleophile with an acid anhydride, an ester and an amide.

Acid anhydride

cont.

Answer

There are two explanations which are commonly given for the order of reactivity observed. One concentrates on the relative reactivities of the C=O group in the four derivatives. The other looks at the leaving groups which are formed following nucleophilic substitution.

Let us look first at the leaving groups which are formed from these four general reactions.

Identify the nucleophiles which are formed after the reaction of a nucleophile with the four carboxylic acid derivatives.

Answer

The departing nucleophiles from the four carboxylic acid derivatives are as follows.

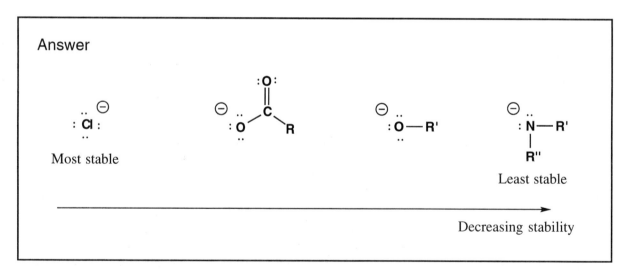

| Chloride, | Alkanoate ion, | Alkoxide, | Alkanamide, |
| from acid chloride | from acid anhydride | from ester | from amide |

The rate of reaction depends on the stability of the nucleophilic leaving groups. If the acid chloride is the most reactive of the acid derivatives, and the amide is the least reactive, what is the order of stability for the leaving groups?

Answer

Most stable   Least stable

⟶

Decreasing stability

The more stable the leaving group, the less difficult it is to remove it from the molecule. The chloride anion must be the best leaving group since acid chlorides are the most reactive of the acid derivatives. In order to be a good leaving group, the chloride anion must be stable and unreactive. One major contributer to the stability of a negative charge is the electronegativity of the atom bearing the charge. Can you suggest how electronegativity affects the stability of the four anions above?

Answer

Chlorine is an electronegative atom on the right of the periodic table and is able to cope with a negative charge much better than atoms to the left of it in the periodic table. Oxygen is more electronegative than nitrogen and so both the alkanoate and alkoxide ions will be more stable than the alkanamide anion.

Electronegativity does not explain the difference in stability between the alkanoate ion and the alkoxide ion since the charge is on oxygen in both ions. We have seen in previous sections that inductive and resonance effects are important in stablising charge. Explain how these effects stabilise or destabilise charge in alkanoate and alkoxide ions.

Answer

The alkanoate ion is more stable than the alkoxide ion since it is possible to spread the charge between two oxygens by resonance. The mechanism by which the charge can be spread in the alkanoate ion is as follows.

Resonance is not possible for the alkoxide ion and so the charge is localised on one oxygen. The alkyl group attached has an inductive effect whereby it pushes electrons towards the oxygen. This serves to intensify the negative charge and destabilise it.

The relative ease of the leaving groups is reflected in their relative basicities. The chloride ion is a very weak base and remains fully ionised in water. The carboxylate ion is a weak base and is in equilibrium with the carboxylic acid in water. The methoxide ion and the alkanamide ion are both strong bases and react with water to give an alcohol and an amine, respectively.

Another explanation for the different rates of reaction between different carboxylic acid derivatives depends on the relative reactivities of the C=O group in these derivatives. Why are the carboxylic acid derivatives reactive to nucleophiles in the first place?

Answer

The carboxylic acid derivatives are electrophilic. The electrophilic centre is the carbonyl carbon atom.

$$X = Cl, OCOR, OR, NR_2$$

Electrophilic centre

Explain why the carbonyl carbon is electrophilic.

Answer

The carbonyl oxygen is electronegative and has a greater share of the electrons in the carbonyl double bond. As a result, the carbonyl bond is polarised such that the oxygen is slightly negative and the carbon is slightly positive.

Note that X is also electronegative and we would expect this to further increase the electrophilicity of the carbonyl carbon.

Explain how different X groups affect the electrophilicity and hence the reactivity of the carbonyl group.

Answer

If X is strongly electronegative (e.g. chlorine), it will withdraw electrons from the carbonyl carbon making that carbon more electrophilic and more reactive to nucleophiles. The more electronegative X is, the greater effect it has on the carbonyl carbon and the greater the reactivity towards nucleophiles. Chlorine is more electronegative than oxygen which is more electronegative than nitrogen.

The electron pulling effect of X on the carbonyl carbon described above is an inductive effect. With amides, there is an important resonance contribution which *decreases* the electrophilicity of the carbonyl–carbon. Identify the electron-rich centres in amides and identify whether any of these electron-rich centres can form a bond to the electrophilic carbonyl carbon.

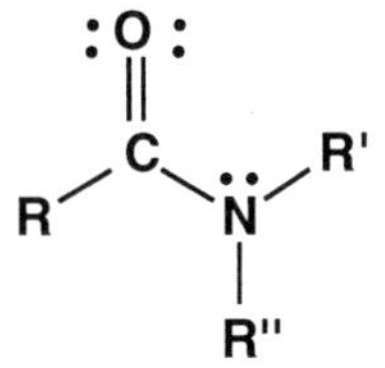

Answer

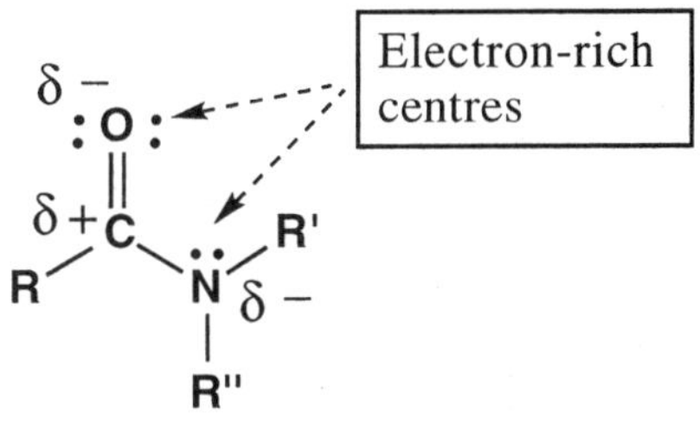

Electron-rich centres

The lone pair of nitrogen can form a π bond to the carbonyl carbon as long the π bond already present breaks.

Show the mechanism by which the lone pair of electrons on nitrogen can form a bond to the carbonyl carbon. Show the resonance structure which would result.

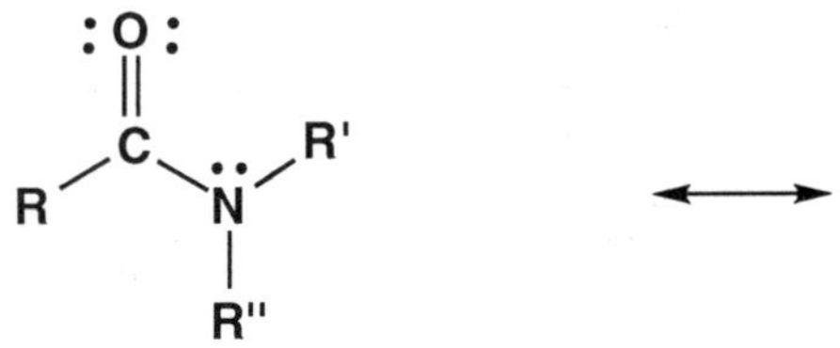

Answer

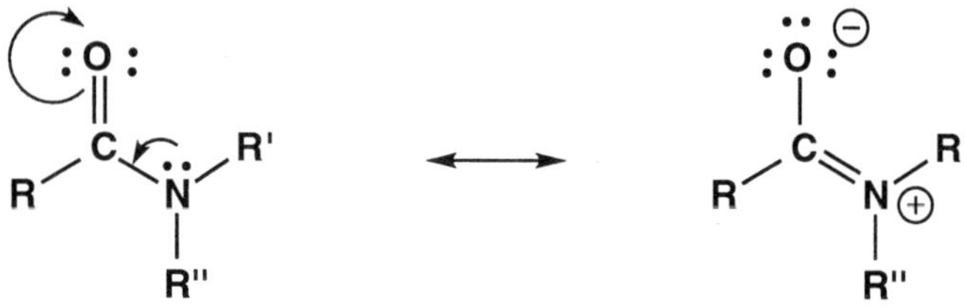

The nitrogen has a lone pair of electrons which are available to form a bond to the neighbouring electrophilic carbonyl carbon. However, carbon can only have a maximum of four bonds, so as the new bond is formed, the weakest bond of the four already present must break. This is the π bond. Both electrons move onto oxygen to give a third lone pair of electrons and a negative charge. The nitrogen atom has used two electrons to form a bond. These electrons are now shared between carbon and nitrogen and so the nitrogen gains a positive charge.

The structures above are resonance structures. How do you think this resonance might lower the reactivity of amides to nucleophiles?

Answer

Resonance 'breaks open' the carbonyl C=O bond so that it is less prone to attack by an incoming nucleophile. The electrophilic character of the carbonyl carbon is also lessened since the electron deficiency is shared between carbon and nitrogen.

This resonance is important in amides but does not occur to such a great extent with esters, acid anhydrides or acid chlorides. Can you suggest why not? Draw the structures which would be formed by such a resonance and identify anything which might be undesirable and therefore unstable about them.

Answer

The products which would be obtained from a similar resonance with the other acid derivatives are the following.

From an acid chloride

From an acid anhydride

From an ester

In these structures the positive charge ends up on an oxygen or a chlorine atom. These atoms are more electronegative than nitrogen and are 'less happy' with a positive charge. These resonance structures might occur to a small extent with esters and acid anhydrides, but are far less likely in acid chlorides. This trend also matches the trend in reactivity.

Resonance is least likely to occur in acid chlorides due to the unfavourable positive charge which would result on the chlorine atom. Therefore, the carbonyl group remains strongly electrophilic and prone to attack by nucleophiles. Resonance may occur to some extent with acid anhydrides and esters, but this will be less significant than with amides, since oxygen is less happy with a positive charge than nitrogen. Can the resonance effect explain why acid anhydrides are more reactive than esters though?

Answer

It can. Acid anhydrides have two carbonyl groups and the nucleophile could attack either. Resonance can take place with either carbonyl group. Therefore, the lone pair of the central oxygen is 'split' between both groups. This means that the carbonyl groups of an acid anhydride are more electrophilic than the carbonyl group of an ester.

We have now seen that the carboxylic acid derivatives have different reactivities and we have gone through the various factors which cause this. We shall now look at the consequences of this as far as the reactions of the carboxylic acid derivatives are concerned. Earlier, we saw that acid chlorides can be used to synthesise other carboxylic acid derivatives by reaction with suitable nucleophiles. Do you think it would be possible to do the reverse reaction and synthesise an acid chloride from an ester, amide or acid anhydride?

Answer

It would not be possible. We would be trying to convert a more stable, less reactive acid derivative into a less stable, more reactive acid chloride.

Taking this into account and remembering the order of reactivity of carboxylic acid derivatives, which of the following reactions would be possible?

(a)

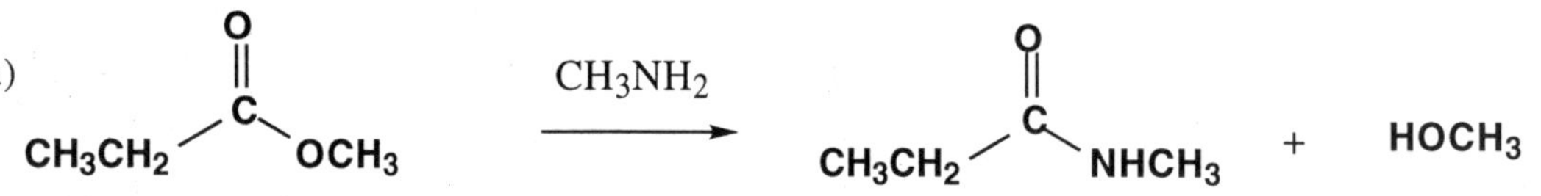

CH_3NH_2

(b)

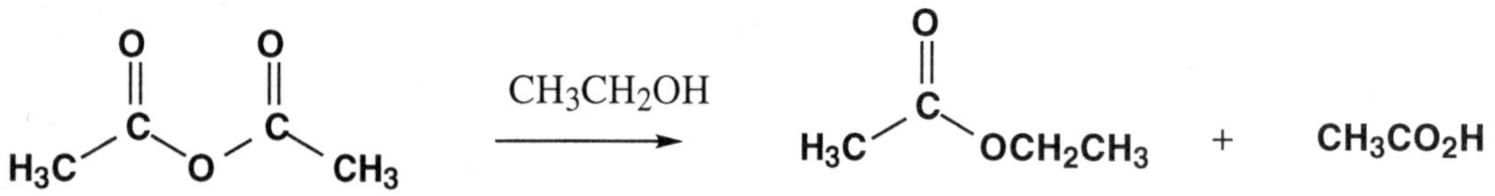

CH_3CH_2OH

(c)

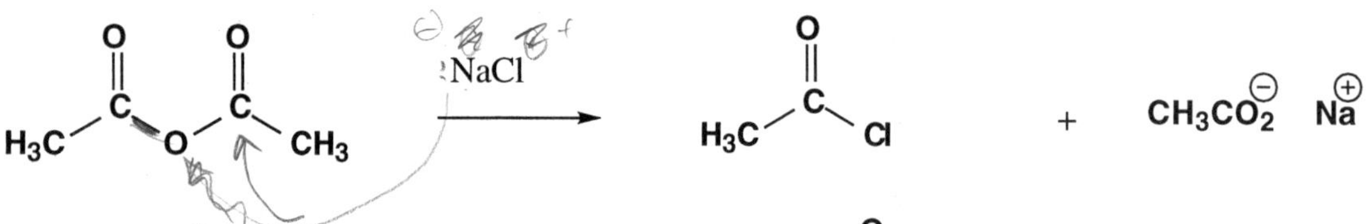

$NaCl$

(d)

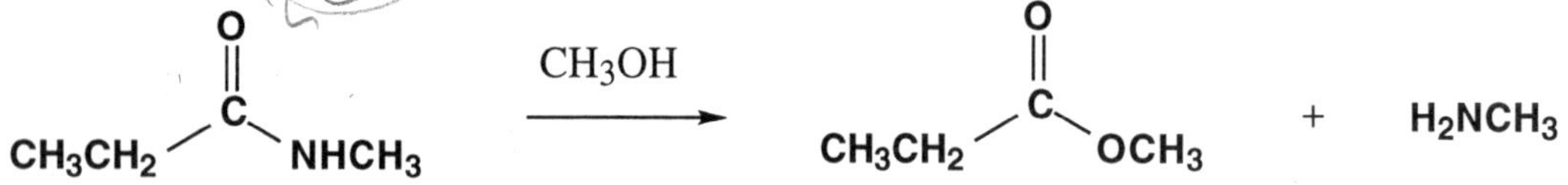

CH_3OH

(e)

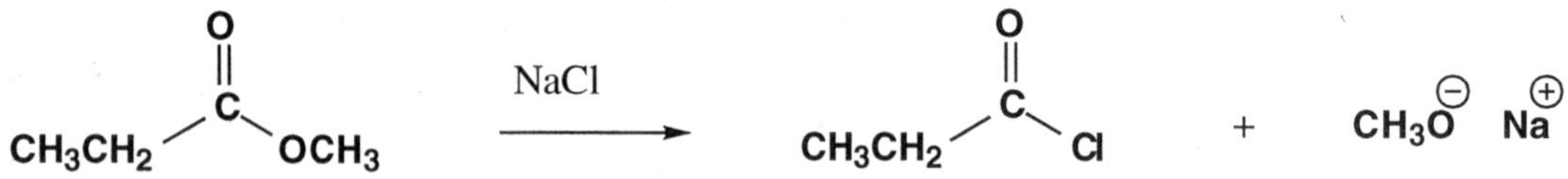

$NaCl$

(f)

$CH_3CO_2^{\ominus}\ Na^{\oplus}$

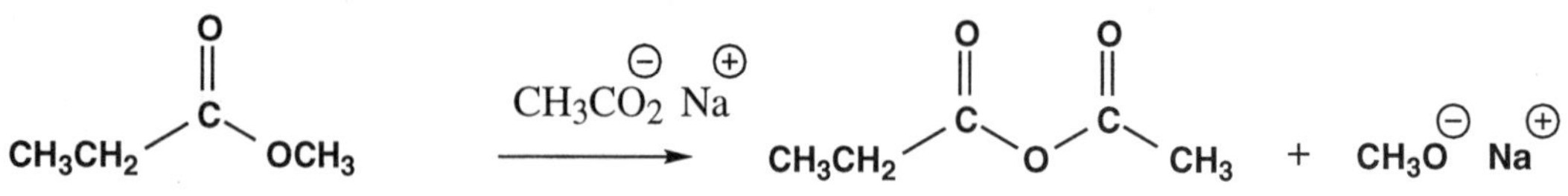

Favoured reactions:

Disfavoured reactions:

Answer

The reactions above represent the following functional group transformations and are favoured or disfavoured as shown.

(a) Ester $\xrightarrow{\text{amine}}$ Amide FAVOURED

(b) Acid anhydride $\xrightarrow{\text{alcohol}}$ Ester FAVOURED

(c) Acid anhydride $\xrightarrow{\text{chloride}}$ Acid chloride DISFAVOURED

(d) Amide $\xrightarrow{\text{alcohol}}$ Ester DISFAVOURED

(e) Ester $\xrightarrow{\text{chloride}}$ Acid chloride DISFAVOURED

(f) Ester $\xrightarrow{\text{acetate}}$ Acid anhydride DISFAVOURED

You should now be able to predict whether the various acid derivatives can react with the following reagents: $RCO_2^- Na^+$; ROH; $RO^- Na^+$; NH_3; RNH_2; R_2NH. List only the *favoured* reactions for cach the carboxylic acid derivatives and identify the functional groups which are formed.

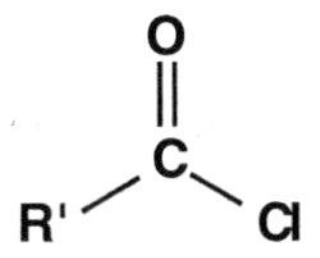

Acid chloride

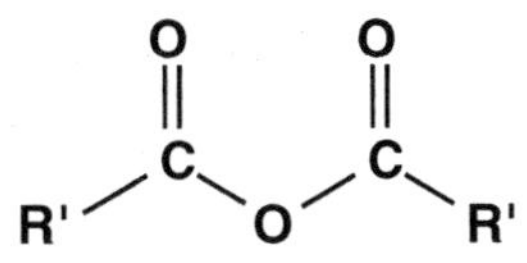

Acid anhydrides

cont.

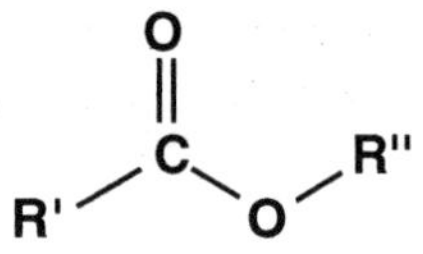

Esters

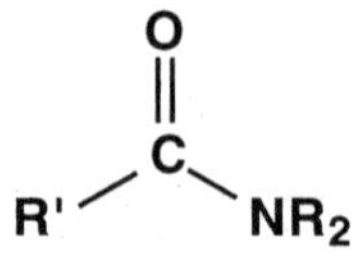

Amides

Answer

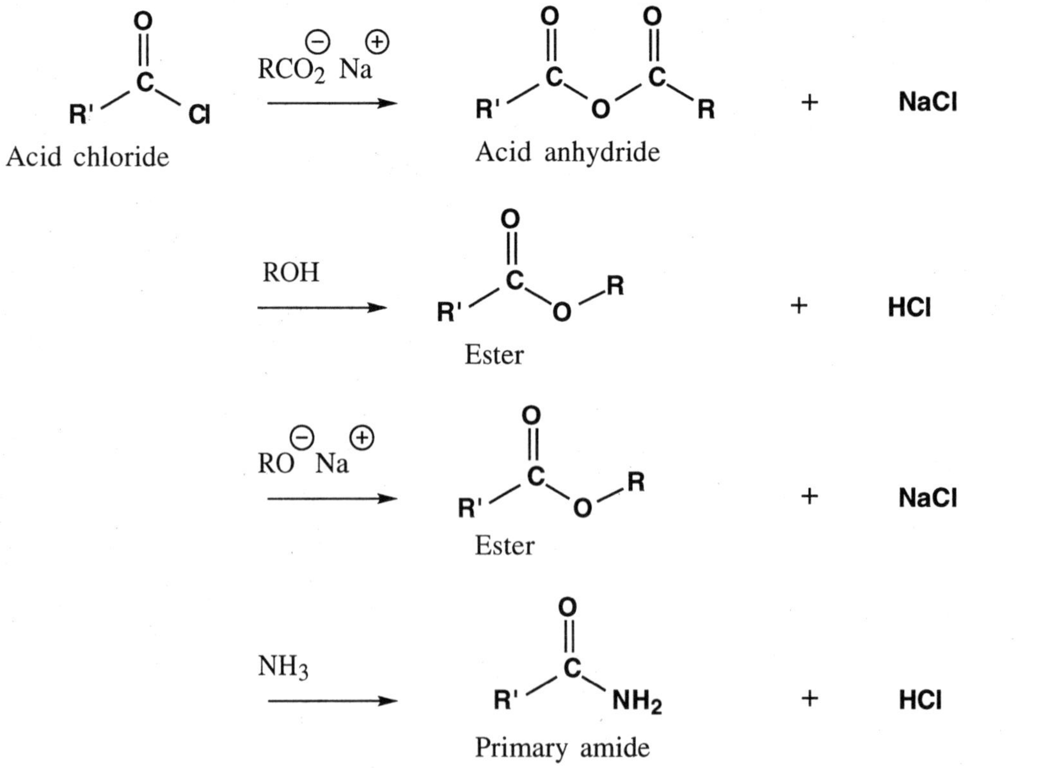

Acid chloride → RCO$_2^-$ Na$^+$ → Acid anhydride + **NaCl**

→ ROH → Ester + **HCl**

→ RO$^-$ Na$^+$ → Ester + **NaCl**

→ NH$_3$ → Primary amide + **HCl**

cont.

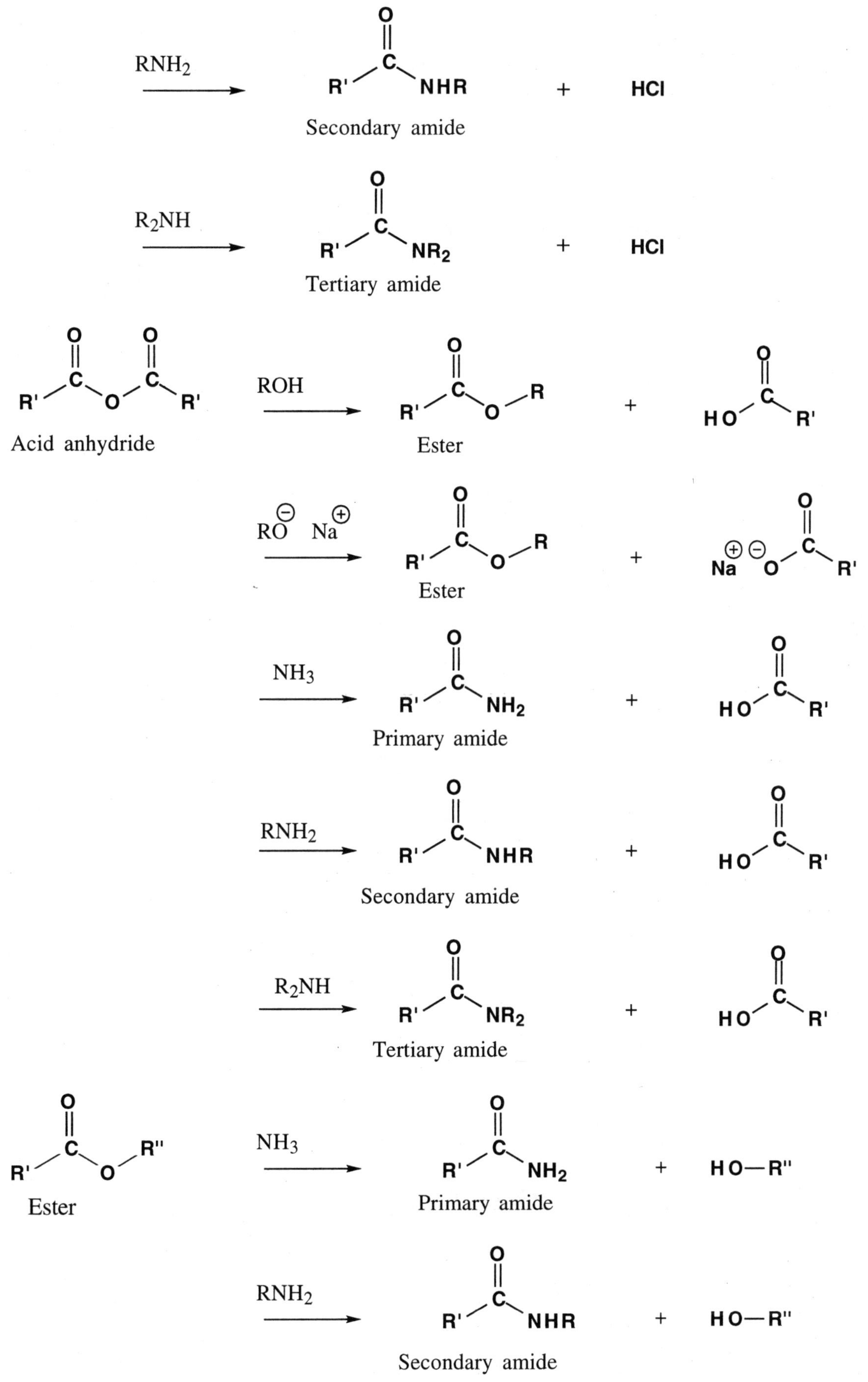

$$\xrightarrow{RNH_2}$$

R'—C(=O)—NHR   +   **HCl**

Secondary amide

$$\xrightarrow{R_2NH}$$

R'—C(=O)—NR₂   +   **HCl**

Tertiary amide

R'—C(=O)—O—C(=O)—R'

Acid anhydride

$$\xrightarrow{ROH}$$

R'—C(=O)—O—R   +   HO—C(=O)—R'

Ester

$$\xrightarrow{RO^{\ominus} \ Na^{\oplus}}$$

R'—C(=O)—O—R   +   Na$^{\oplus}$ $^{\ominus}$O—C(=O)—R'

Ester

$$\xrightarrow{NH_3}$$

R'—C(=O)—NH₂   +   HO—C(=O)—R'

Primary amide

$$\xrightarrow{RNH_2}$$

R'—C(=O)—NHR   +   HO—C(=O)—R'

Secondary amide

$$\xrightarrow{R_2NH}$$

R'—C(=O)—NR₂   +   HO—C(=O)—R'

Tertiary amide

R'—C(=O)—O—R''

Ester

$$\xrightarrow{NH_3}$$

R'—C(=O)—NH₂   +   HO—R''

Primary amide

$$\xrightarrow{RNH_2}$$

R'—C(=O)—NHR   +   HO—R''

Secondary amide

cont.

$$\xrightarrow{R_2NH} \quad \underset{\text{Tertiary amide}}{R' - \overset{\overset{\displaystyle O}{\|}}{C} - NR_2} \quad + \quad HO - R''$$

9.7 Nucleophilic substitution with the hydroxide ion—hydrolysis

Carboxylic acid derivatives react with aqueous sodium hydroxide. Identify the nucleophilic centre in sodium hydroxide and identify the electrons which will be used for the new bond.

Answer

Sodium hydroxide is a salt and when dissolved in water will give sodium ions and hydroxide ions. The nucleophile will be the negatively charged hydroxide ion. The nucleophilic centre will be the oxygen and the electrons for the new bond will come from one of the three lone pairs on the oxygen.

$$:\overset{\ominus}{\underset{..}{\ddot{O}}} - H \qquad \text{Hydroxide ion}$$

Show the reaction you would expect for the nucleophilic substitution of the following acid chloride with sodium hydroxide.

$$H_3C - \overset{\overset{\displaystyle :O:}{\|}}{C} - \overset{..}{\underset{..}{Cl}}: \quad \xrightarrow{\text{NaOH}}$$

Answer

The general equation for an acid chloride undergoing nucleophilic substitution is as follows.

$$H_3C - \overset{\overset{\displaystyle :O:}{\|}}{C} - \overset{..}{\underset{..}{Cl}}: \quad + \quad \overset{\ominus}{:Nu} \quad \longrightarrow \quad H_3C - \overset{\overset{\displaystyle :O:}{\|}}{C} - Nu \quad + \quad :\overset{..}{\underset{..}{Cl}}:^{\ominus}$$

cont.

The equation for this problem will therefore be as follows.

(Sodium acts as a counterion on both sides of the equation.)

This is not the final product of this reaction, however. Look carefully at the carboxylic acid which is obtained. Identify the nucleophilic and electrophilic centres and suggest what might happen next, bearing in mind that the reaction is being carried out in basic conditions.

Answer

Nucleophilic substitution generates a carboxylic acid which has an acidic proton. The reaction of the hydroxide ion with a carboxylic acid produces a carboxylate ion.

Show the mechanism for the above reaction.

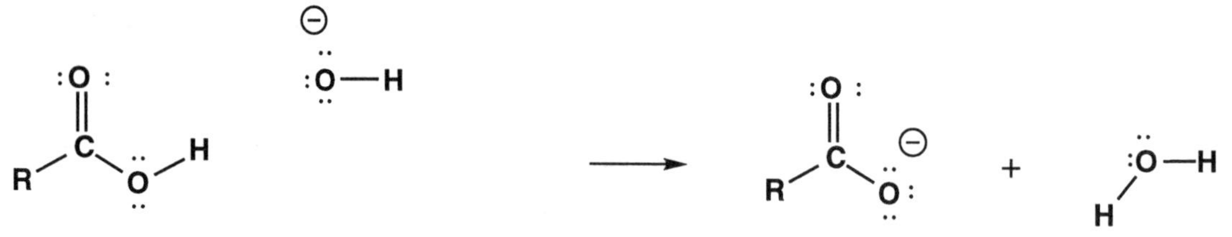

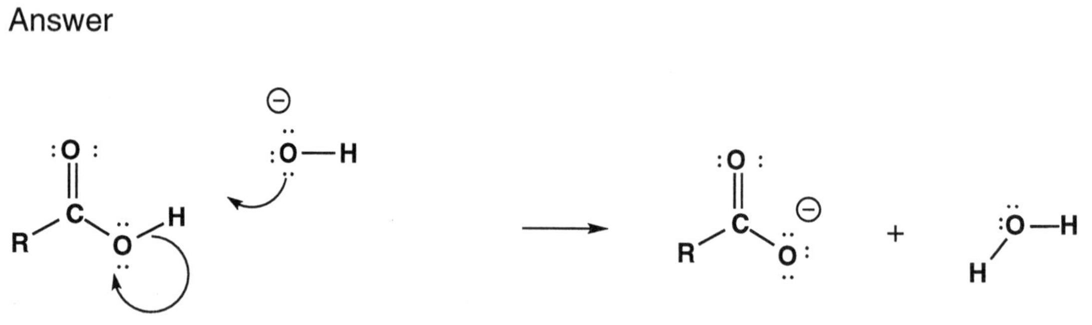

Answer

All the carboxylic acid derivatives react with the hydroxide ion and can be hydrolysed back to the original carboxylic acid. Heating may be required for amides, which is another indication of the difficulty of reacting amides with nucleophiles. Write the general equations for the reactions of acid chlorides, acid anhydrides, esters and amides with sodium hydroxide.

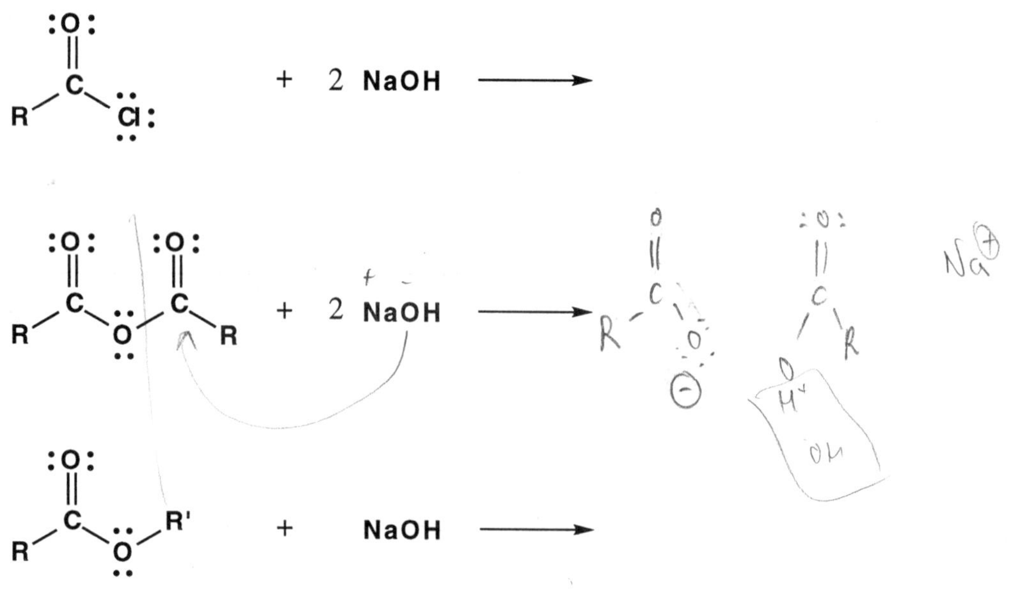

cont.

Answer

Note that the reaction of the ester gives the alcohol as the other product and not the alkoxide ion. The alkoxide ion is a strong base and picks up a proton from water to generate the hydroxide ion.

The same is true for the reaction of an amide. The leaving group here is the strongly basic alkanamide ion. This too picks up a proton from water to give the amine as the other product plus the hydroxide ion.

Both the alkoxide ion and the alkanamide ion are stronger bases than the hydroxide ion and as such are poorer leaving groups. The reason these reactions are possible is the formation of the stable carboxylate salt as the final product. This is an irreversible reaction which drives the reaction through to completion.

In all these hydrolysis reactions the final product is the carboxylate salt which is soluble in water. Suppose you wanted the carboxylic acid instead and not the salt. How would you get it?

Answer

You would have to add aqueous acid (e.g. dil. HCl) to the aqueous solution. The acid would protonate the carboxylate salt to give the carboxylic acid. The free carboxylic acid would no longer be soluble in aqueous solution and would precipitate out as a solid or as an oil.

Draw a mechanism to show how an acid protonates the carboxylate ion to give the carboxylic acid.

Answer

A common error which students make is to draw the mechanism for protonation as follows.

Why is this wrong?

Answer

Curly arrows show the movement of electrons and can only show how a nucleophile attacks an electrophile and not the other way round. The arrow must start from a source of electrons (i.e. a lone pair or a bond). A proton has no electrons and therefore it is meaningless to draw an arrow from a proton.

9.8 Hydrolysis (acid catalysed)

In Section 9.7, we saw the hydrolysis of carboxylic acid derivatives with a hydroxide ion to give carboxylic acids. It is also possible to hydrolyse carboxylic acid derivatives with aqueous acid.

Once again, this is a nucleophilic substitution, but what is the nucleophile? Is it the hydroxide ion?

Answer

Since acid conditions are being used, the concentration of hydroxide ion is very low. The nucleophile in acid conditions must be water itself.

What is the nucleophilic centre in water and what would be the product if water reacted with an electrophile (E$^+$)?

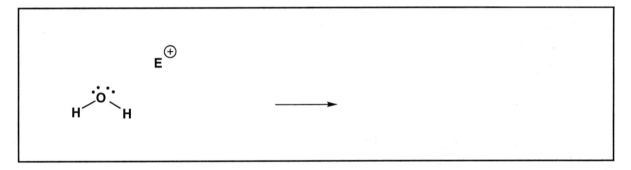

Answer

The nucleophilic centre is oxygen and it would use one of its lone pairs of electrons to form a bond to the electrophile.

Would you expect water to be a better nucleophile than the hydroxide ion? Give a reason for your answer.

Answer

Water would be a far poorer nucleophile than the hydroxide ion. The hydroxide ion is negatively charged and is neutralised when it forms a new bond to an electrophile. Water on the other hand would gain a positive charge when it forms a bond to an electrophile.

Since water is a poor nucleophile, it needs 'help' in order to hydrolyse carboxylic acid derivatives. That help comes from the acid which catalyses the reaction. This is the same sort of catalysis which we saw in Section 8.4 for acid-catalysed nucleophilic additions to aldehydes and ketones.

The same arguments holds here. In order to act as a catalyst, the acidic proton provided by the acid must take part in the reaction. Identify whether the proton is an electrophile or a nucleophile, then identify possible sites where it could react with the following ester.

Answer

The proton is positively charged and is an electrophile. It will therefore react with a nucleophilic centre on the ester. There are two nucleophilic centres on the ester—the two oxygen atoms.

There are two nucleophilic centres on the ester. Draw the mechanisms and products for each of these centres when they react with a proton.

Answer

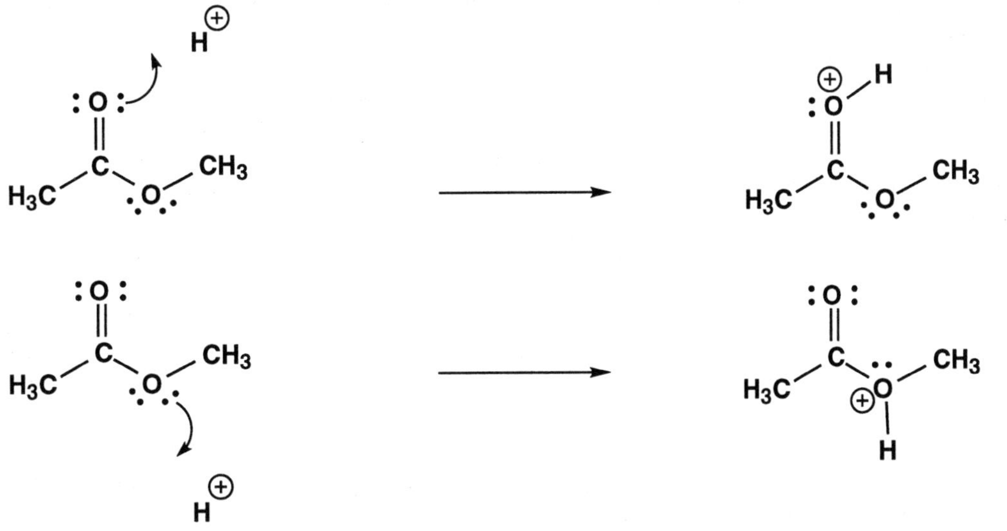

Both of these are likely, but only one of them is going to help catalyse nucleophilic substitution by water. Remember the first step in nucleophilic substitution involves the addition of the nucleophile to the carbonyl carbon and the breaking of the π bond. Which of the protonated intermediates above is most likely to assist this step and why?

Answer

The intermediate which is protonated on the carbonyl oxygen is more likely to assist the first step of the reaction mechanism. The mechanism requires the π bond to break with both electrons moving onto the carbonyl oxygen. If this oxygen is protonated, it is positively charged and the mechanism is driven to neutralise that charge.

Now that we have identified the protonated intermediate which will catalyse the first step, draw the first two steps of the hydrolysis mechanism.

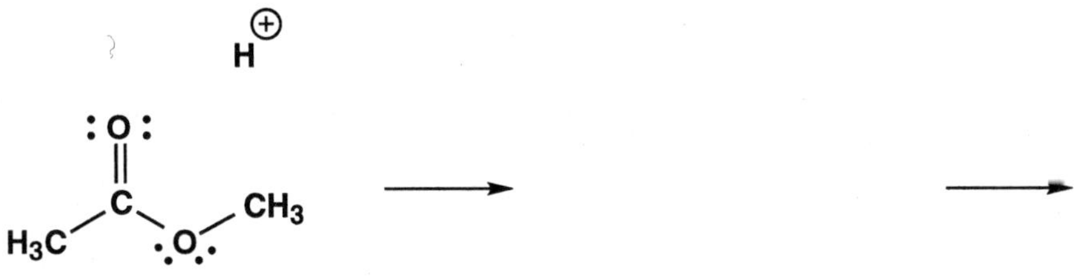

Answer

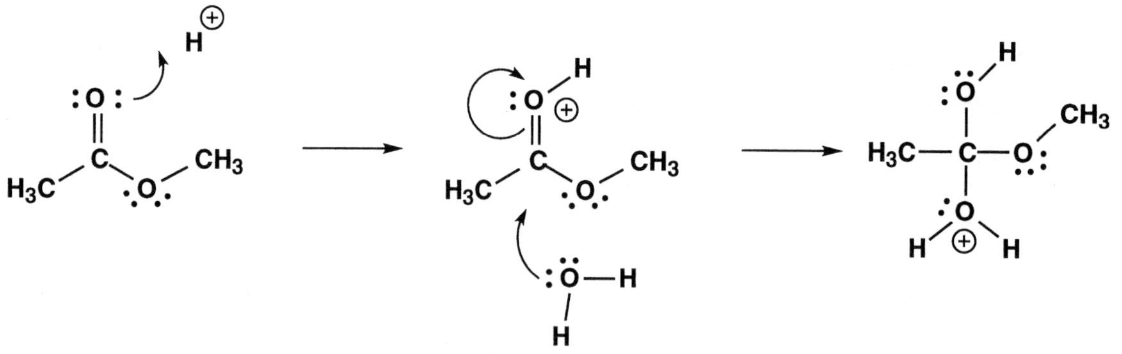

We have shown how acid catalysis encourages the weakly nucleophilic water to react.

The intermediate now has a positive charge on oxygen. That is not particularly stable, but we can neutralise the charge by losing the attached proton. By losing the proton, the oxygen can gain both electrons in that bond and lose its charge.

Draw a mechanism to show this and draw the intermediate which is formed.

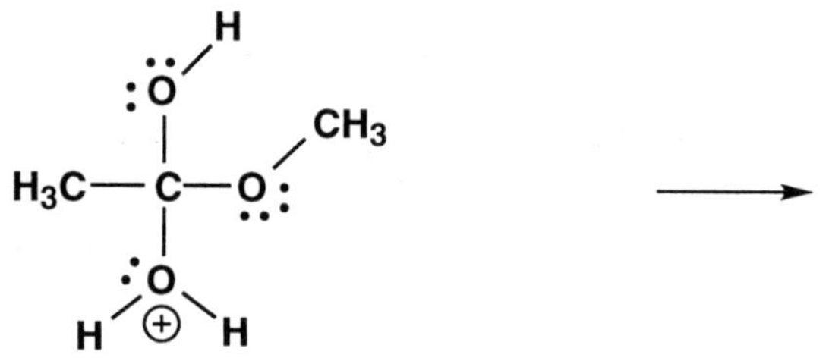

Answer

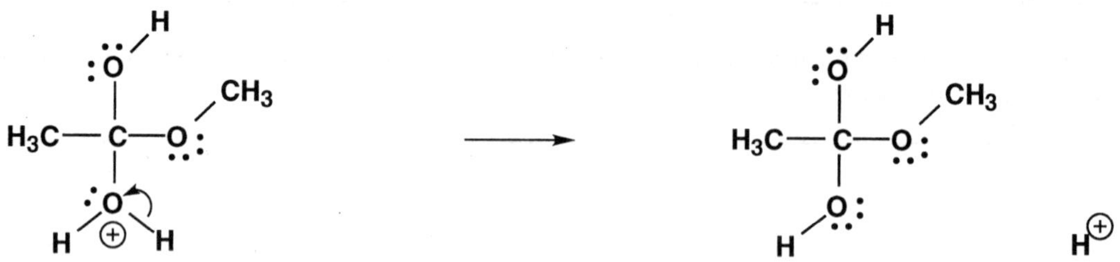

We are now at the stage where we want to reform the carbonyl group and lose the methoxy group. There are, however, a couple of problems. Draw the mechanism as you would have drawn it for the uncatalysed mechanism using a lone pair of electrons from the oxygen to reform the π bond. Draw the products which would be obtained.

Answer

There are a couple of things which make this mechanism difficult. One is to do with the structure which is formed and one is to do with the leaving group. Suggest what the problems might be.

Answer

The carbonyl oxygen has to donate a pair of electrons to reform the π bond. This means that the oxygen will gain a positive charge which is not very favourable. Therefore, the oxygen is weakly nucleophilic and this makes the mechanism more difficult.

As far as the leaving group is concerned, it is negatively charged, and it is also a strong base. It is not possible to have this as a leaving group under acid conditions.

We could solve both these problems with a better leaving group. If the leaving group was stable in acid conditions, it would leave more readily and this would make up for the weakly nucleophilic oxygen which has to provide the electrons for the π bond.

Once again, acid catalysis can assist the mechanism. If we react the intermediate with a proton before reforming the π bond, the mechanism will be much better. How many different nucleophilic centres could a proton react with on the intermediate shown?

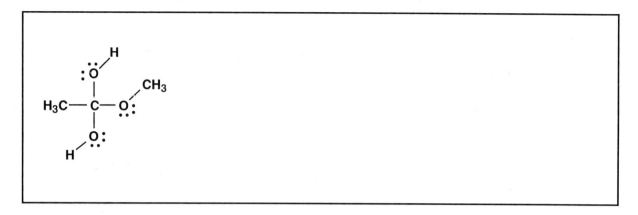

Answer

There are three possible nucleophilic sites—the three oxygen atoms.

Draw the intermediates which would be obtained if the proton reacted with each of these sites.

Answer

I   II   III

Each of these intermediates is likely, but only one of them will help the mechanism go more easily. Suggest which one. (Look again at the bonds which have to be made and the bonds which have to be broken, then identify which intermediate will help this process.)

Answer

Intermediate II will assist the mechanism. When the C–O bond breaks, the electrons in the bond will both move onto a positively charged oxygen. This will neutralise the charge and give methanol. Methanol is a much better leaving group than the methoxide ion. The better leaving group will make up for the disadvantage of having a weakly nucleophilic oxygen.

Draw the next two stages of the mechanism to show protonation, then the departure of the leaving group.

Answer

There is one more step to put in to complete the mechanism and that is to lose the proton from the carbonyl oxygen. Draw a mechanism for that step.

Answer

We have now gone through all the stages of the mechanism. Draw the complete mechanism below.

Answer

Note how the leaving groups are written above the various 'stage arrows' of the mechanism so that they don't clutter up the overall mechanism. The same mechanism is involved in the acid-catalysed hydrolysis of acid chlorides, acid anhydrides and amides. Draw the mechanism for the acid-catalysed hydrolysis of the following amide

Answer

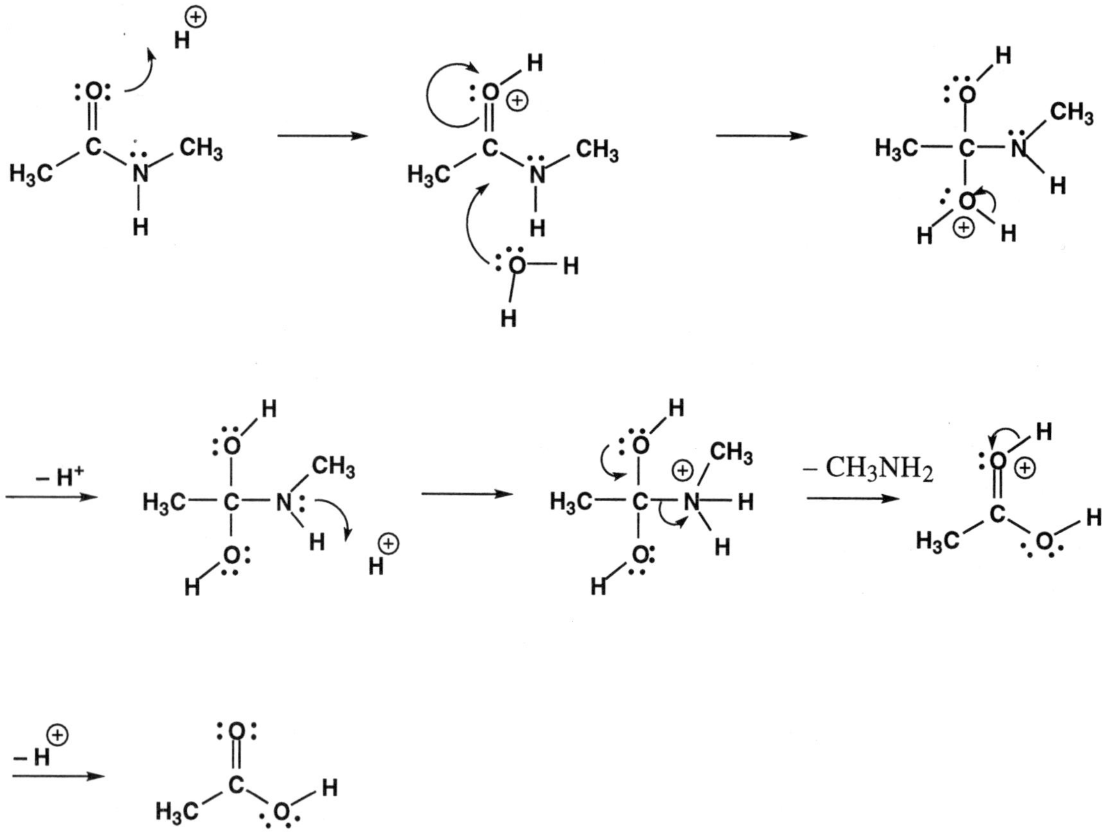

There is one extra detail which we have to consider with the acid hydrolysis of amides. The leaving group in the above mechanism is methylamine (CH_3NH_2). Would you expect this to react further in any way? Remember the reaction is being carried out under acid conditions.

Answer

Since methylamine is a base and the reaction is being carried out under acidic conditions, the amine will become protonated and the final product will be the ion ($CH_3NH_3^+$).

Taking this into account, how much acid would be required for the acid hydrolysis of amides compared with other acid derivatives such as esters?

Answer

With an ester, you only need a catalytic amount of acid since the protons used during the reaction mechanism are regenerated. With an amide, the resulting amine which is formed will form a bond to a proton and 'mop it up'. Therefore, you would need at least 1 equivalent of acid to allow for that.

Summary

- Acid chlorides, acid anhydrides, esters and amides undergo nucleophilic substitution reactions with nucleophiles.

- Acid chlorides are more reactive than acid anhydrides. Acid anhydrides are more reactive than esters and esters are more reactive than amides.

- The rate of reaction is determined by the stability of the leaving group and the electrophilic character of the carbonyl carbon.

- Chloride ion is a better leaving group than a carboxylate anion. The carboxylate ion is a better leaving group than the alkoxide ion and the alkoxide ion is a better leaving group than an alkanamide ion.

- The electrophilic character of the carbonyl carbon is increased by strong electronegative groups such as chlorine.

- The electrophilic character of the carbonyl carbon is decreased by any resonance involving the carbonyl group.

- Ketones and aldehydes cannot undergo nucleophilic substitution since there is no good leaving group.

- Carboxylic acid derivatives undergo nucleophilic substitution rather than nucleophilic addition since the products obtained are more stable than those which would be formed from nucleophilic addition.

- Hydrolysis of carboxylic acid derivatives can be carried out under basic or acidic conditions.

Nucleophilic Substitutions of Alkyl Halides

10.1 Nucleophilic substitution of primary alkyl halides

We shall consider the reaction between methyl iodide and the hydroxide ion as a simple example of nucleophilic substitution involving a primary alkyl halide.

$$I—CH_3 \quad + \quad HO^{\ominus} \quad \longrightarrow \quad HO—CH_3 \quad + \quad I^{\ominus}$$

Iodomethane
(methyl iodide)

Methanol

Identify the nucleophilic and electrophilic centres of both iodomethane and the hydroxide ion and explain why they are nucleophilic or electrophilic.

Answer

The hydroxide ion is nucleophilic because it is negatively charged. The negative charge is on the oxygen, so this is the nucleophilic centre.

$$:\overset{\ominus}{\underset{..}{O}}—H$$

Iodomethane has an electronegative iodine atom joined to a less electronegative carbon atom. The iodine will therefore demand a greater share of the electrons in the carbon–iodine bond. As a result, the iodine will be slightly negative and the carbon will be slightly positive.

cont.

$$\overset{\delta-}{:\ddot{I}}\overset{\delta+}{-}\overset{\overset{\displaystyle H}{\diagup}}{\underset{\underset{\displaystyle H}{\vDash}}{C}}\blacktriangleleft H$$

Therefore, the carbon will be an electrophilic centre and the iodine a weakly nucleophilic centre.

Taking into account the nucleophilic and electrophilic centres in each molecule, which atoms will be involved in the formation of a new bond if a reaction is to take place?

Answer

An electron-rich or nucleophilic centre will provide the electrons to form a bond to an electron-deficient or electrophilic centre. Therefore, the oxygen on the hydroxide ion will provide the electrons for the new bond to the electrophilic carbon on methyl iodide.

If the oxygen of the hydroxide ion is providing both electrons for the new bond, which electrons will it use?

Answer

The oxygen will use one of its lone pairs of electrons to form the new bond.

If a new bond is formed to the alkyl halide's electrophilic carbon and nothing else happens, how many bonds will that carbon have? Is this possible? If not, what must happen?

$$:\ddot{I}-\overset{\overset{\displaystyle H}{\diagup}}{\underset{\underset{\displaystyle H}{\vDash}}{C}}\blacktriangleleft H \qquad \longleftarrow\!-\!-\!-\!-\!-\!- \qquad \text{New bond forming}$$

Answer

If a new bond is formed to the carbon of iodomethane and nothing else happens, the carbon will have five bonds. This is not possible and so one of the original four bonds must break at the same time as the new bond is formed.

Which bond is the most likely one to break and where will the electrons go?

Answer

There are three C–H bonds and one C–I bond. C–H bonds are strong and therefore the most likely bond to break is the C–I bond. Both electrons will move onto the iodine to give a fourth lone pair of electrons and a negative charge. Iodine is electronegative and can stabilise this charge. If the C–H bond had broken it would have generated an unstable, highly reactive hydride anion. This would have required far more energy. Therefore, the C–I bond is broken in preference to the C–H bond.

Taking all the above points into consideration, draw a mechanism showing what will happen when the hydroxide ion forms a bond to iodomethane.

Answer

The lone pair from oxygen is used to form the new bond to carbon at the same time as the carbon–iodine bond is broken. Both electrons in the old C–I bond move onto iodine to give a fourth lone pair of electrons.

cont.

Note how the mechanistic arrow is drawn from the hydroxide ion. It starts from the source of electrons—the lone pair. It points to where the middle of the new bond will be formed. Some textbooks may show the arrow pointing directly to the electrophilic carbon of the alkyl halide (see below), but this is not formally correct. Arrows should only point to atoms if the electrons are going to end up as a lone pair on that atom.

Correct

Incorrect

There are two more points about this mechanism which need to be made. First of all, you will notice that the hydroxide ion approaches iodomethane from one side while the iodide leaves from the opposite side. Can you suggest why the mechanism has been drawn in this way?

Answer

(1) Both the hydroxide and the iodide are negatively charged and will repel each other so it makes sense that they are as far apart as possible.

(2) The hydroxide ion has to find room between the substituents in order to reach the carbon atom. There is more room to attack from the rear since the large iodine atom blocks approach from the front.

There is another interesting feature of this mechanism and that concerns the other three substituents on the carbon (the three hydrogens in this case). Study the mechanism above and suggest what happens to these substituents.

Answer

Both the starting iodide and the alcohol product are tetrahedral compounds. In both molecules, the three hydrogens form an 'umbrella' shape with the carbon. However, the 'umbrella' is pointing in a different direction in the alcohol product, compared to the alkyl halide starting material. This mean that the 'umbrella' has been turned inside out during the mechanism. Putting this into more scientific language, we say that the carbon centre has been 'inverted'.

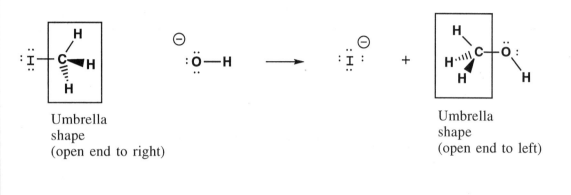

Umbrella
shape
(open end to right)

Umbrella
shape
(open end to left)

This inversion is known as the 'Walden Inversion' and the mechanism as a whole is known as the S_N2 mechanism. The S_N stands for 'substitution nucleophilic'. The 2 signifies that the rate of reaction is second order (bimolecular) and depends on both the concentration of nucleophile and the concentration of alkyl halide. This is because the reaction mechanism involves both molecules in the rate determining step. The S_N2 mechanism holds for the nucleophilic substitutions of primary alkyl halides such as iodomethane.

In the above reaction we used the hydroxide ion as the nucleophile, but hydroxide ions do not come in isolation. There has to be a counterion. For example, sodium hydroxide can be used as a source of hydroxide ions. Give the full reaction equation for iodomethane reacting with sodium hydroxide.

Answer

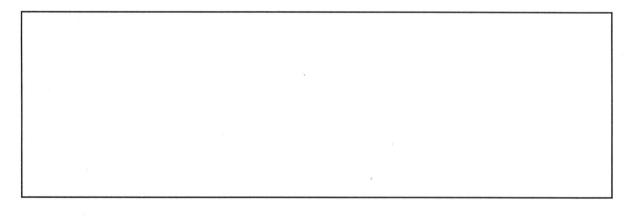

The following salts can be used to provide nucleophiles for nucleophilic substitution. What are the nucleophiles in each of these salts? Show all the lone pairs for these nucleophiles.

(a) NaI (b) NaBr (c) KCN (d) $NaOCH_2CH_3$

(empty box)

Answer

The nucleophiles are as follows.

(a) $: \overset{..}{\underset{..}{I}} :^{\ominus}$ (b) $: \overset{..}{\underset{..}{Br}} :^{\ominus}$ (c) $^{\ominus} : C \equiv N :$ (d) $^{\ominus} : \overset{..}{\underset{..}{O}} — CH_2CH_3$

Give the full reaction equations for the following reactions.

(a)

$I — CH_2CH_3$ + $NaOH$ $\longrightarrow$

(b)

$Br — CH_2CH(CH_3)_2$ + $NaOH$ $\longrightarrow$

(c)

$Br — CH_2CH_3$ + NaI $\longrightarrow$

(d)

$I — CH_2CH_2CH_3$ + KCN $\longrightarrow$

(e)

$I — CH_3$ + $NaOCH_2CH_3$ $\longrightarrow$

Answer

The alkyl groups in the original alkyl halides are boxed to clarify the changes.

(a) $I —\boxed{CH_2CH_3}$ + $NaOH$ $\longrightarrow$ $HO —\boxed{CH_2CH_3}$ + NaI

cont.

(b) Br—CH₂CH(CH₃)₂ + NaOH ⟶ HO—CH₂CH(CH₃)₂ + NaBr

(c) Br—CH₂CH₃ + NaI ⟶ I—CH₂CH₃ + NaBr

(d) I—CH₂CH₂CH₃ + KCN ⟶ NC—CH₂CH₂CH₃ + KI

(e) I—CH₃ + NaOCH₂CH₃ ⟶ CH₃CH₂O—CH₃ + NaI

Neutral nucleophiles can also react with alkyl halides by nucleophilic substitution. Ammonia and alkylamines are examples of neutral nucleophiles. What is the nucleophilic centre in these molecules?

Answer

The nitrogen atom is the nucleophilic centre for ammonia and amines since the nitrogen is more electronegative than carbon and has a lone pair of electrons.

If ammonia forms a bond to an alkyl halide, which electrons are used and what effect has this on the nitrogen atom?

Answer

The nitrogen uses its lone pair of electrons to form the new bond. As a result, the nitrogen will effectively lose an electron and will gain a positive charge.

Draw the mechanism for the reaction of ammonia with 1-iodopropane.

Answer

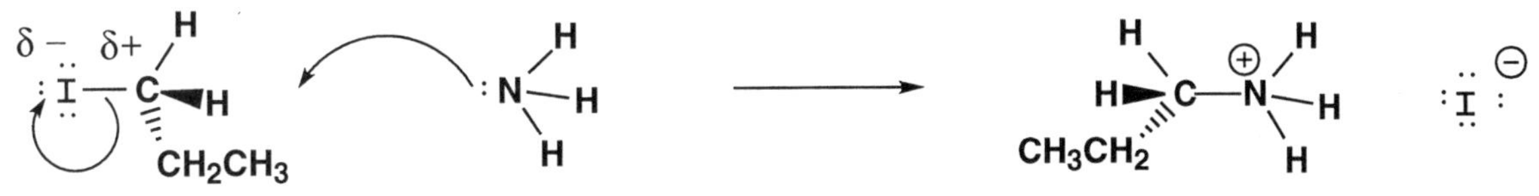

The lone pair on the nitrogen is used to form the bond to the electrophilic carbon of the alkyl halide. At the same time as this bond is being formed, the bond between carbon and iodine is broken with both electrons moving onto the iodine to form a fourth lone pair of electrons. Iodine leaves as the iodide ion. The remaining substituents are inverted during the mechanism. Since the nitrogen has used up its lone pair of electrons in a new bond, it effectively loses an electron and gains a positive charge. The iodide leaving group acts as the counterion and an alkylammonium salt is obtained.

The reaction of an amine with an alkyl halide is a nucleophilic substitution as far as the alkyl halide is concerned. The same reaction can be called an alkylation from the amine's point of view. This is because the amine has gained an alkyl group from the reaction.

Reaction of an amine with an alkyl halide gives an alkylammonium salt, but the free base can be obtained by reaction with sodium hydroxide. The free base can be extracted from the aqueous solution by using an organic solvent such as diethyl ether or ethyl acetate.

$$CH_3CH_2CH_2-\overset{\oplus}{N}H_3 \quad \overset{\ominus}{I} \; + \; NaOH \longrightarrow CH_3CH_2CH_2-NH_2 \; + \; NaI \; + \; H_2O$$

Salt (soluble in water) — Free base (soluble in organic solvent) — NaI (soluble in water)

Draw a mechanism to show the formation of the free base.

Answer

Mechanism

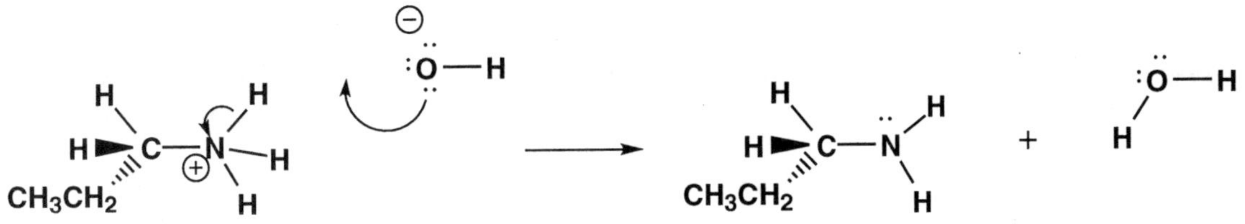

What products would you get from the following reactions?

(a) $CH_3 - I$ $\xrightarrow{CH_3CH_2 - NH_2}$ $\xrightarrow{NaOH}$

(b) $CH_3CH_2 - I$ $\xrightarrow{CH_3 - NH_2}$ $\xrightarrow{NaOH}$

Answer

(a) $CH_3 - I + CH_3CH_2 - NH_2 \longrightarrow CH_3CH_2 - \overset{\oplus}{N}H_2 - CH_3$ $\overset{\ominus}{I}$ $\xrightarrow{NaOH}$ $CH_3CH_2 - NH - CH_3$

$+ NaI + H_2O$

(b) $CH_3CH_2 - I + CH_3 - NH_2 \longrightarrow CH_3 - \overset{\oplus}{N}H_2 - CH_2CH_3$ $\overset{\ominus}{I}$ $\xrightarrow{NaOH}$ $CH_3 - NH - CH_2CH_3$

$+ NaI + H_2O$

The same product is obtained from each of the above reactions. Usually there will be more than one way of synthesising a particular compound. One of the skills required by an organic chemist is to identify which of several possible synthetic routes might be the most effective route to use.

10.2 Nucleophilic substitution of secondary and tertiary alkyl halides

So far we have looked at nucleophilic substitutions of primary alkyl halides. We shall now look at nucleophilic substitutions of secondary and tertiary halides.

Let us consider the following iodides. Identify these iodides as primary, secondary or tertiary.

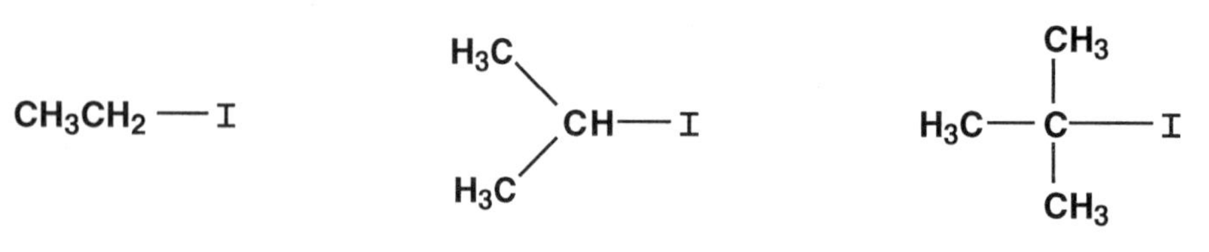

$CH_3CH_2 - I$

$\underset{H_3C}{\overset{H_3C}{\diagdown}} CH - I$

$H_3C - \underset{CH_3}{\overset{CH_3}{\overset{|}{\underset{|}{C}}}} - I$

Answer

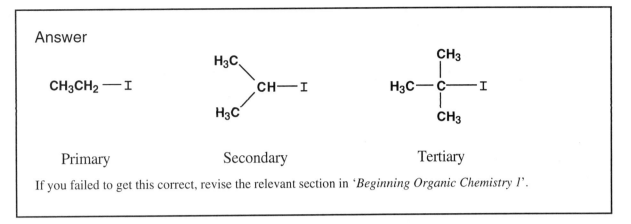

$$CH_3CH_2 — I$$

Primary

Secondary

Tertiary

If you failed to get this correct, revise the relevant section in *'Beginning Organic Chemistry 1'*.

Identify the electrophilic centre in each of these molecules.

Answer

The electrophilic centre in all these iodides is the carbon to which the iodine atom is attached. The iodine is electronegative and has a greater share of the electrons in the carbon–iodine bond. This results in a polar bond. The iodine has a slightly negative charge and the carbon has a slightly positive charge. Consequently, the carbon atom will be electron 'deficient' and will be electrophilic.

You have seen that primary alkyl halides react with nucleophiles by nucleophilic substitution and that the attacking nucleophile forms a bond to the electrophilic carbon of the alkyl halide at the same time as the bond between the carbon and the halide is broken. Both electrons in the C–X bond move onto the halogen atom to form a halide ion which leaves the molecule. The mechanism is repeated below for the reaction of iodomethane with the hydroxide ion.

Remember that the incoming nucleophile attacks the electrophilic carbon from the rear of the molecule. Consequently, the carbon centre is inverted.

The primary, secondary and tertiary alkyl halides which we looked at earlier are redrawn below to show the tetrahedral shape of the electrophilic carbon centre.

Study these molecules and if possible make models of them. Do you think they will all react equally well with the hydroxide ion by the mechanism above? Give reasons for your answer.

Answer

You would not expect these molecules to react equally well with the hydroxide ion by the mechanism shown. You would expect iodomethane to react fastest and the tertiary iodide to react slowest. You may have given two possible explanations for this—one electronic and one steric.

First of all, you may have noticed that the different iodoalkanes have a different number of alkyl groups attached to the electrophilic centre. As you have learnt already, alkyl groups have an electron-donating effect which tends to counterbalance a neighbouring electrophilic centre. The electron-donating effect means that the neighbouring electrophilic centre will be less electron-deficient and less reactive to nucleophiles.

The second reason is steric. A methyl group is a bulky group compared with a hydrogen and can therefore act like a shield against any approaching nucleophile. Three such shields makes it very difficult for a nucleophile to reach the electrophilic carbon of a tertiary alkyl halide.

Primary alkyl halide. Electrophilic carbon is easily accessible.

Secondary alkyl halide. Electrophilic carbon is a 'bit of a squeeze'.

Tertiary alkyl halide. Electrophilic carbon is inaccessible.

10.3 Nucleophilic substitution of tertiary alkyl halides by the S_N1 mechanism

The steric bulk of alkyl substituents makes it very difficult for a nucleophile to reach the electrophilic carbon centre of tertiary alkyl halides. Therefore, these compounds should not react with nucleophiles. However, in practice they do. For example, the reaction of the hydroxide ion with the tertiary halide below gives an alcohol product.

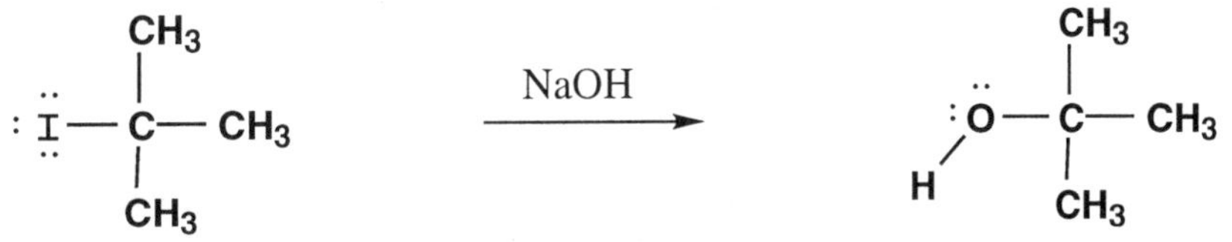

Nucleophilic substitution has taken place, but this time it is found that the rate of reaction depends only on the concentration of the alkyl halide. Clearly, the nucleophile has to be present if the reaction is to take place at all, but it does not matter whether there is 1 equivalent of the nucleophile or an excess. The rate is unaffected. Since the reaction rate only depends on the alkyl halide, the mechanism is known as the S_N1 reaction, where S_N stands for Substitution Nucleophilic. The 1 shows that the reaction is first order and that only one of the reactants affects the reaction rate. What is this S_N1 mechanism? Let us look again at the problem which faced us with the S_N2 mechanism. The attacking nucleophile was unable to reach the electrophilic carbon centre of tertiary alkyl halides because of the bulky alkyl substituents blocking access. How then could the carbon centre become more accessible?

If the alkyl groups are not to block the incoming nucleophile, they will have to move further apart from each other. However, this is not possible as long as the carbon centre is tetrahedral. If, however, the C–I bond is broken, we would get a flat carbocation as shown below.

Draw a mechanism to show how the C–I bond could be broken and explain where the electrons in the bond end up.

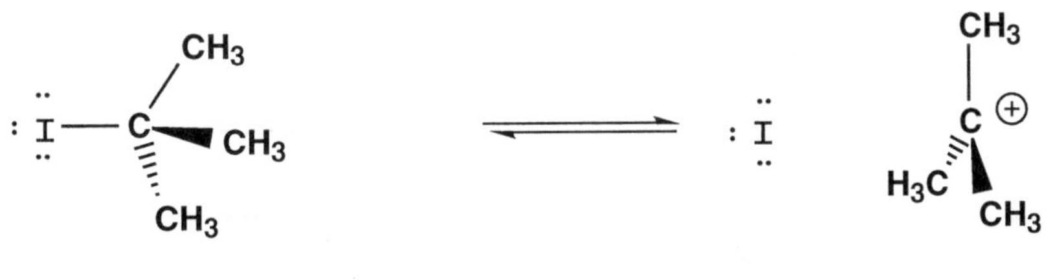

Answer

The C–I bond breaks and both electrons in the bond move onto the iodine atom to give a fourth lone pair of electrons on iodine. Iodine has effectively 'gained' an electron since it now has both electrons from the bond to itself, instead of sharing them with carbon. Therefore, iodine gains a negative charge.

The carbocation is planar instead of tetrahedral. Can you suggest the advantages of this?

Answer

The ion is more stable when it is planar since all three alkyl groups are as far apart from each other as possible. The carbon atom is now sp^2 hybridised with an empty $2p_y$ orbital.

We have suggested that the C–I bond breaks to form a planar carbocation. However, bonds don't normally break without reason. In this case, there are a couple of driving forces which make the process possible. One of these is steric. Compare the position of the alkyl groups in the tetrahedral alkyl halide and in the carbocation and suggest what steric advantage the carbocation has over the tetrahedral alkyl halide.

Answer

The alkyl groups are crowded together in the tetrahedral structure. In the carbocation, they are further apart and there is less steric strain as a result.

The other reason is electronic. There are two electronic factors which are involved in stabilising the intermediate carbocation and hence encouraging the reaction. One of these is an inductive effect. Explain with the aid of diagrams how inductive effects stabilise the carbocation.

Answer

We have seen in previous sections that alkyl groups can have an inductive effect whereby they 'push' electrons to a neighbouring centre. This electron-donating effect could be seen as encouraging the departure of the halide ion. More importantly, the electron-donating effect of the alkyl groups help to stabilise the carbocation since the inductive effect reduces and hence stabilises the positive charge.

→	Inductive effect encourages departure of halide.

→	Inductive effect stabilises carbonium ion.

There is another method by which the alkyl groups can stabilise the carbocation above. Identify what it is, referring back to Section 5 if you have forgotten.

Answer

The above carbocation is also stabilised by hyperconjugation. We discussed hyperconjugation in Section 5. Hyperconjugation is possible if neighbouring C–H σ bonds can overlap with the empty 2p orbital of the carbocation. The overlap of orbitals allows the positive charge to be spread over the molecule and helps to stabilise it.

Identify the C–H σ bonds which can take part in hyperconjugation of the carbocation below.

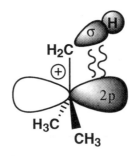

Answer

All the C–H bonds above are available for hyperconjugation.

Draw a diagram showing how the empty p orbital of the carbocation can overlap with one of the C–H σ bonds.

Answer

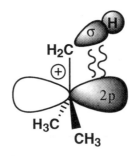

Although we have shown the favourable reasons for the formation of the carbocation, it is important to realise that the ion still has a positive charge and is therefore less stable than the uncharged tetrahedral structure. The iodide ion can depart but will come back again to reform the C–I bond. In other words, there is an equilibrium between the tetrahedral structure and the carbocation. Draw a mechanism to show the iodide reforming a bond to the carbocation and explain which species is the nucleophile and which is the electrophile.

Answer

The iodide ion has a negative charge and is the nucleophile. The carbocation has a positive charge and is the electrophile. The iodide ion uses one of its lone pairs of electrons to form a new bond to the electrophilic centre of the carbocation. The resulting structure is tetrahedral rather than planar.

We have shown the iodide ion acting as a nucleophile to give back the starting material, but if there is another nucleophile present, it too could react with the carbocation. You should now be able to write the complete S_N1 mechanism for the following reaction.

$$I-C(CH_3)_3 \xrightarrow{\text{NaOH}} HO-C(CH_3)_3$$

Answer

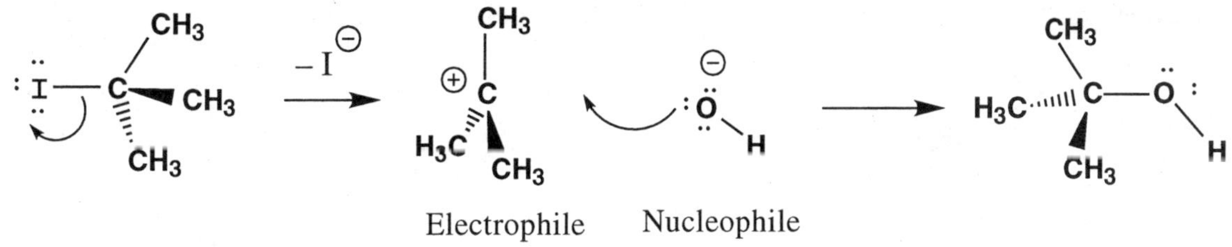

Electrophile   Nucleophile

cont.

Stage 1

The C–I bond breaks first with both electrons moving onto the iodine to give an iodide ion. The carbocation is stabilised by the electron-donating effect of three alkyl groups as well as hyperconjugation. Steric strain is also relieved. This is the rate determining step.

Stage 2

The hydroxide oxygen has a negative charge and is therefore a nucleophilic centre. A lone pair of electrons is used to form a bond to the electrophilic centre of the carbocation. The tetrahedral centre is reformed.

In the mechanism above, we have shown the hydroxide ion coming in from the right-hand side. Is there any reason why the hydroxide ion can not come in from the left hand side instead?

Answer

No reason at all. The carbocation is flat so the hydroxide ion can attack equally well from one side or the other.

Since the incoming nucleophile can attack from either side of the carbocation, there is no overall inversion of the carbon centre in the S_N1 mechanism. With a molecule such as the one above, this is not important. However, it *is* important when the reaction is carried out on chiral molecules.

Identify the asymmetric centre in the following alkyl halide.

Answer

* = asymmetric centre

Assign the asymmetric centre as (*R*) or (*S*), showing how you assign the asymmetric centre. (If you have forgotten how to do this revise the section on optical isomers in *'Beginning Organic Chemistry 1'*.)

Answer

The asymmetric centre has three carbons and one iodine attached to it. The iodine has the highest atomic number and takes priority 1.

In order to determine the priority of the alkyl groups, we move to the next highest atom. In the case of the methyl groups this has to be a hydrogen, whereas it is a carbon in the other two alkyl groups. Therefore, the methyl group has the lowest priority (priority 4).

By similar arguments, the ethyl substituent has a lower priority than the isopropyl substituent.

Connecting groups 1–3 is anticlockwise. Configuration is (*S*).

Show the mechanism for the reaction of this iodide with the hydroxide ion.

Answer

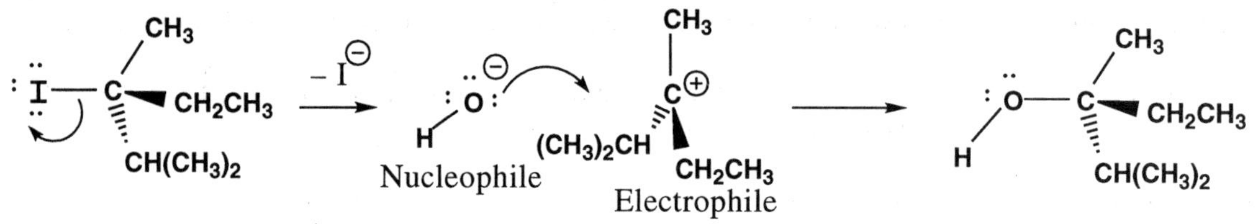

The mechanism shows the hydroxide ion attacking from the left-hand side, but it can also attack from the right-hand side, to give the alternative product.

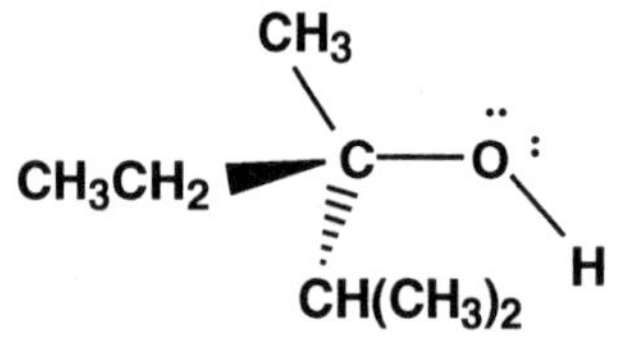

Compare the two possible products and explain whether they are identical molecules or not.

Answer

The two alcohols formed in the reaction are mirror images of each other. They are not superimposable and are therefore configurational isomers.

What is the term for two non-superimposable mirror images such as these?

Answer

They are enantiomers.

Starting from the optically active (*S*) isomer of the alkyl halide, would you expect the product alcohol to have optical activity?

Answer

You would not. The reaction has resulted in racemisation of the asymmetric centre leading to equal quantities of both enantiomers in the product*. A racemic mixture such as this would not have optical activity.

(* However, it is has to be noted that **total** racemisation does **not** occur for most S_N1 reactions. The reasons for this are not within the scope of this text.)

If this same reaction had gone by an S_N2 mechanism would you have expected the product to have optical activity?

Answer

You would. The S_N2 mechanism is stereospecific. The nucleophile would have attacked from one side only and inverted the asymmetric centre to give the product as a single enantiomer.

Consider the following alkyl iodide. Identify whether it is chiral or not and identify whether it is a primary, secondary or tertiary halide.

Answer

The alkyl iodide is chiral. The asymmetric centre is marked with an asterisk. It is also a tertiary alkyl halide.

State whether you think nucleophilic substitution with a hydroxide ion is more likely to go by an S_N1 or S_N2 mechanism.

Answer

Since the iodide is tertiary, the reaction is more likely to go by an S_N1 mechanism.

State whether you would expect the product to be a single enantiomer or a racemate.

Answer

This question is slightly sneaky. Although this reaction goes by the S_N1 mechanism, you would expect the product to be a single enantiomer and not a racemate. That is because the asymmetric centre is not the reaction centre and will not be affected by the reaction.

10.4 Nucleophilic substitution of secondary alkyl halides

It is generally fair to say that nucleophilic substitution of primary alkyl halides will take place via the S_N2 mechanism whereas nucleophilic substitution of tertiary alkyl halides will take place by the S_N1 mechanism. What about secondary alkyl halides? Do they react by the S_N1 or the S_N2 mechanism?

Answer

It is impossible to predict with any certainty whether a secondary alkyl halide will react by the S_N1 or the S_N2 mechanism. If the alkyl groups are small, there is more chance that the mechanism will be S_N2. If the alkyl groups are large, there is more chance that the mechanism will be S_N1.

That being the case, how could you find out experimentally whether the following reaction is S_N1 or S_N2? Look back to remind yourself what S_N1 and S_N2 stand for.

$$\text{I} —\text{CH(CH}_3)_2 \quad + \quad \textbf{NaOH} \quad \longrightarrow \quad \textbf{H O}—\textbf{CH(CH}_3)_2 \quad + \quad \textbf{NaI}$$

Answer

The only way to find out for certain whether the reaction is S_N1 or S_N2 is to try out the reaction and see whether the reaction rate depends on the concentration of both reactants (S_N2) or whether it depends on the concentration of the alkyl halide alone (S_N1).

There is another way in which you could tell whether the following reaction is S_N1 or S_N2. Suggest what that might be.

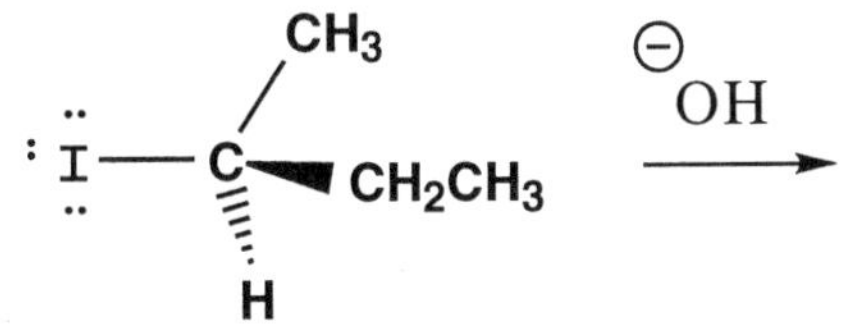

Answer

The alkyl halide has four different substituents at the reaction centre. It is chiral and will have optical activity. If a pure enantiomer is used in the reaction and the reaction goes by an S_N1 mechanism, then a mixture of enantiomers will be obtained with little or no optical activity.

On the other hand, if the S_N2 mechanism is followed, the asymmetric centre will be inverted and a single enantiomeric product will be obtained which *will* have optical activity.

Measuring the optical activity of the product and comparing that value with the literature value will tell you whether one or other or both of these mechanisms have been followed.

Summary

- Alkyl halides can undergo nucleophilic substitutions with nucleophiles.

- Primary alkyl halides undergo nucleophilic substitution by the S_N2 mechanism.

- Tertiary alkyl halides undergo nucleophilic substitution by the S_N1 mechanism.

- Secondary alkyl halides undergo nucleophilic substitution by either the S_N1 or the S_N2 mechanism.

- S_N stands for Substitution Nucleophilic.

- The rate of reaction in an S_N2 mechanism is affected by the concentration of both reactants.

- The rate of reaction in an S_N1 mechanism is affected by the concentration of alkyl halide alone.

- The S_N2 mechanism is a concerted mechanism whereby the departure of the outgoing nucleophile coincides with the arrival of the incoming nucleophile. The incoming nucleophile approaches from the rear of the molecule and the carbon centre is inverted.

- The S_N1 mechanism goes through a planar carbocation intermediate. There is no overall inversion.

SECTION 11

Reactions Involving Acidic Protons Attached to Carbon: Aldehydes and Ketones

Reactions Involving Acidic Protons Attached to Carbon: Aldehydes and Ketones

11.1 Acidic protons attached to carbon—aldehydes and ketones

In Section 4 we looked at acidic protons attached to heteroatoms such as halogen, oxygen and nitrogen. Protons attached to carbon are not normally acidic, but in some cases they are. One such situation occurs with aldehydes or ketones when there is a CHR_2, CH_2R or CH_3 group next to the carbonyl group.

Which of the following ketones or aldehydes fit this description?

Answer

Structures (a) and (d) fit this description. Structure (a) has a CH_3 group next to the carbonyl functional group while structure (d) has a CHR_2 group as a neighbour.

The carbon next to any functional group is known as the alpha (α) carbon and any protons attached to it are known as α protons. α protons next to the carbonyl group of a ketone or aldehyde will tend to be acidic.

You should now be able to identify acidic protons in aldehydes or ketones by looking for protons at the α position, but why should these protons be acidic? Let us look at what will happen if a nucleophile like the hydroxide ion acts like a base and forms a bond to one of the acidic α protons in ethanal (acetaldehyde). Draw the mechanism for this and see what structure you end up with. Remember that hydrogen is only allowed one bond.

Answer

A lone pair on the hydroxide oxygen is used to form a new bond to the α proton on ethanal. At the same time as this happens, the C–H bond must break. Both electrons of that bond end up on the carbon atom and give it a negative charge.

The negative charge is on a carbon atom and so the ion is known as a carbanion. For this acid–base reaction to take place, the carbanion must be stabilised in some way. An electronegative atom can stabilise a negative charge. Is carbon electronegative?

Answer

No. Carbon is not electronegative.

Since a carbon atom is not particularly electronegative, the negative charge in the carbanion must be stabilised in some other way. If the charge could be delocalised by resonance, that would help to stabilise it. Is resonance possible in this case? Remember the requirement for resonance is a π bond next to the charged atom, i.e.

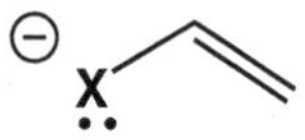

Is such a system present in the carbanion below. Highlight it if it is.

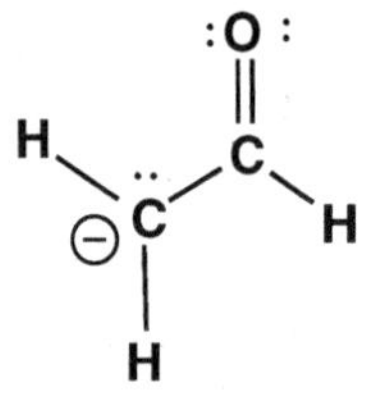

Answer

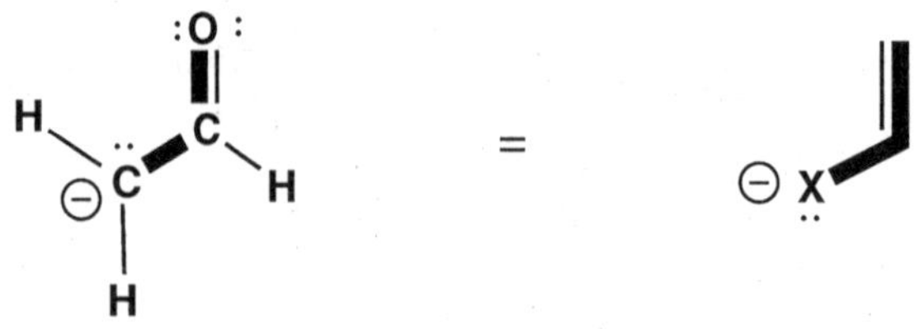

Resonance is possible for this structure. Draw the other resonance structure and the mechanism involved.

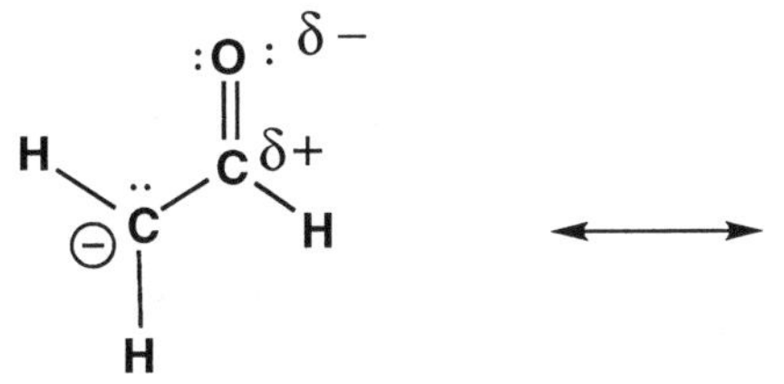

Answer

The negative charge of the carbanion can be stabilised by resonance as shown below. The lone pair of electrons on the carbon are next to the electrophilic carbon of the carbonyl group and are drawn in to form a bond between the two carbons. As the bond is formed, the weak π bond of the carbonyl group breaks and both electrons move onto the oxygen. The overall result is that the negative charge ends up on the electronegative oxygen where it is more stable. This structure is known as an enolate ion.

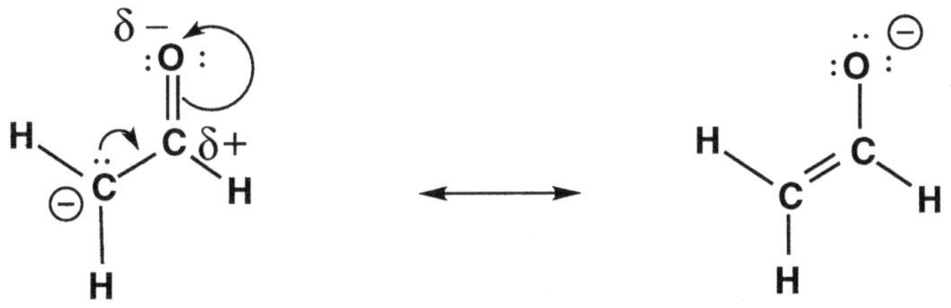

Enolate ion

The mechanism leading to the second resonance structure is exactly the same as the one we looked at for the carboxylate ion.

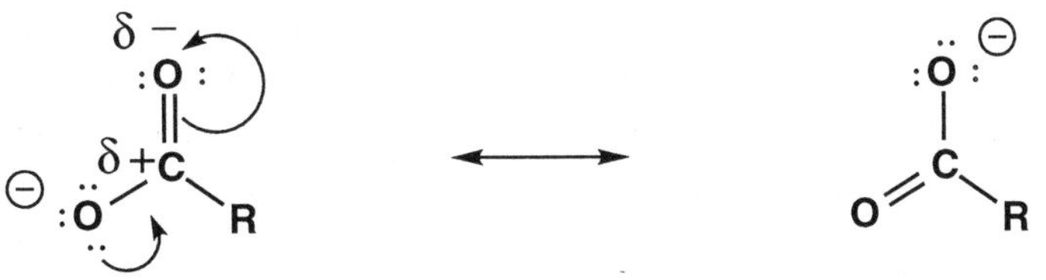

However, whereas both resonance structures are equally stable in the carboxylate ion, the enolate ion is favoured over the carbanion since the negative charge ends up on the electronegative oxygen atom. The enolate ion is preferred to such an extent that a better mechanism would be to show the enolate ion being formed as the acidic proton is lost. Can you devise such a mechanism?

Answer

The improved mechanism would be as follows and gives the more stable enolate ion at the same time as the acidic proton is 'picked off' by the hydroxide ion.

As the hydroxide forms its bond to the acidic proton, the C–H bond breaks. The electrons in that bond could move onto the carbon but the electrophilic carbonyl group 'pulls them in' so that they form a π bond to the carbonyl carbon atom. At the same time, the carbonyl π bond breaks such that both electrons move onto oxygen.

When we write the mechanism in this way, we can see that it is the electronegative oxygen which is really responsible for making the α proton acidic, since it allows the formation of the stable enolate ion.

This mechanism is only possible for hydrogens at the alpha position. Hydrogens at other positions will not be acidic, since formation of a stable enolate ion is not possible; for example neither of the following mechanisms are possible. The charge is localised on one carbon and cannot be delocalised by resonance. It will therefore be too unstable, and the ion cannot be formed.

False Mechanism 1

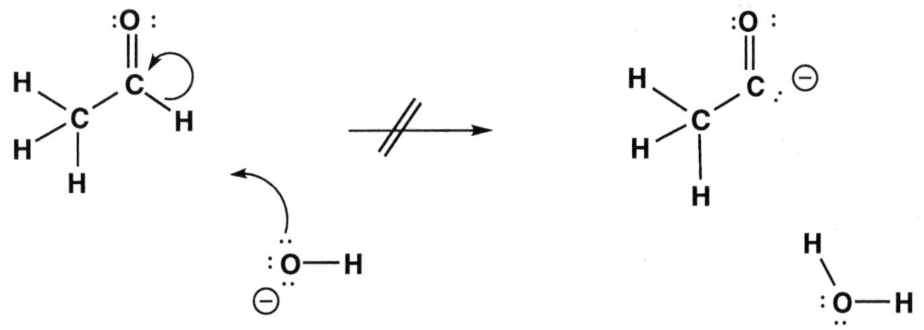

False Mechanism 2

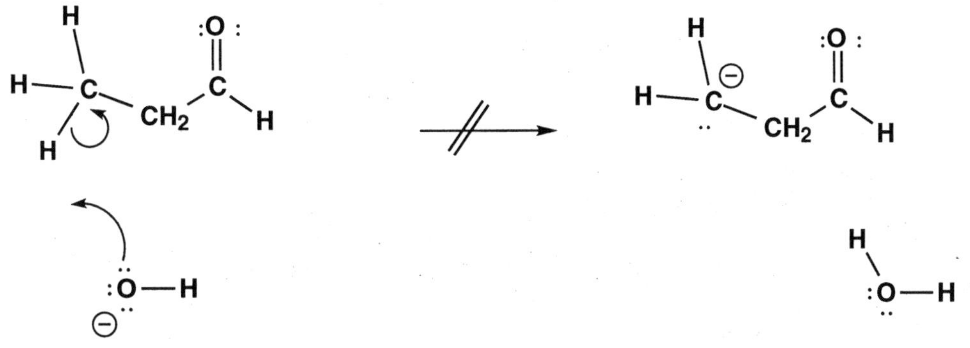

A student disagreed with this statement and drew the following mechanism to try and delocalise the charge on the first of the anions above. What is wrong with this mechanism?

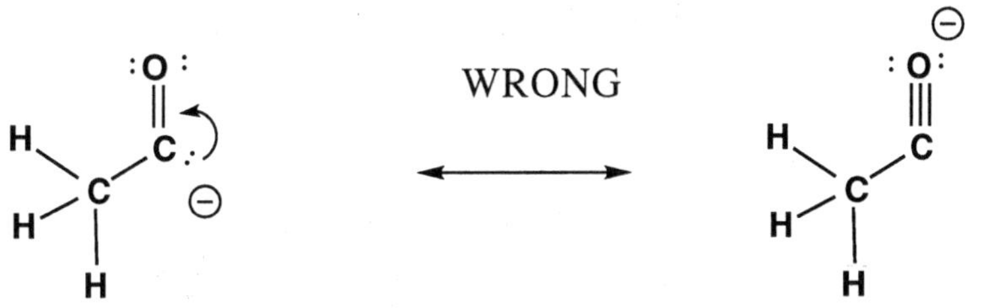

WRONG

Answer

This mechanism is impossible. The student has drawn an arrow to show that the lone pair of electrons on carbon is forming a third bond to oxygen. However, oxygen already has a full octet of valence electrons with two lone pairs and two bonds. It cannot accept another bond.

Therefore, only protons in the alpha position in relation to carbonyl groups can be acidic, since this is the only position where the resulting charge can be delocalised to give the stable enolate ion.

We shall now look at the enolate ion in more detail.

The diagram below shows the two resonance structures for the enolate ion. Resonance structures represent the extreme possibilites for a particular molecule and the true structure is really a hybrid of these resonance structures.

Draw a structure which would show the enolate ion as a hybrid of these two resonance structures.

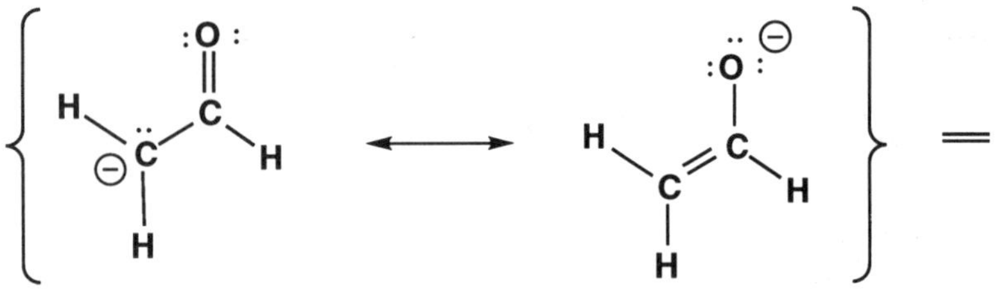

Answer

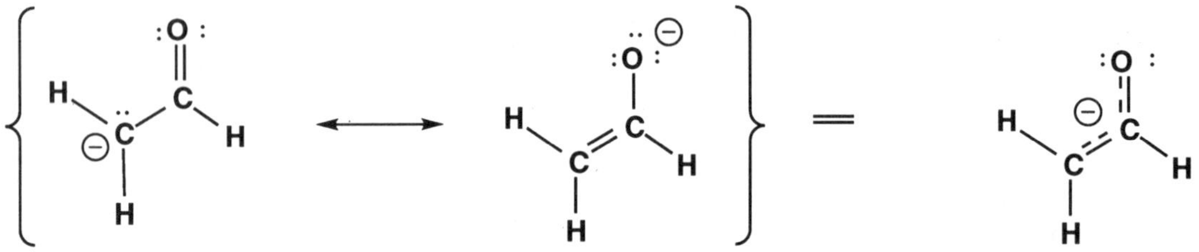

The 'hybrid' structure shows that the negative charge is 'smeared' between three sp² hybridised atoms.

Since these atoms are sp² hybridised, they will be planar and will each have a 2p orbital. Draw an orbital diagram showing how these 2p orbitals could interact to spread the charge between the three atoms.

Answer

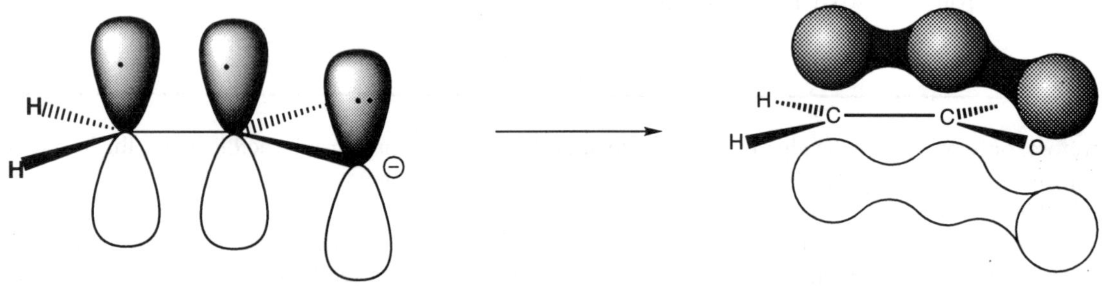

Bearing in mind the interactions above, which of the following methyl hydrogens is the most likely to be lost in the formation of an enolate ion and why?

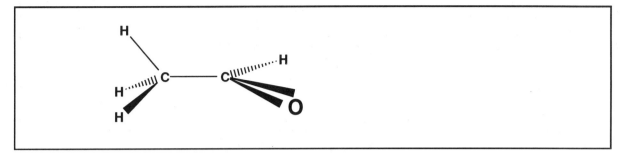

Answer

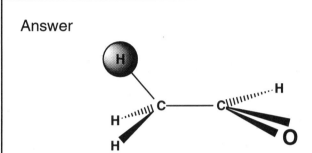

The hydrogen circled is the one which will be lost since the σ C–H bond is correctly orientated to interact with the π orbital of the carbonyl bond.

Draw a diagram to show this interaction.

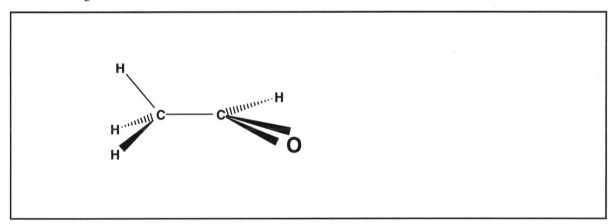

Answer

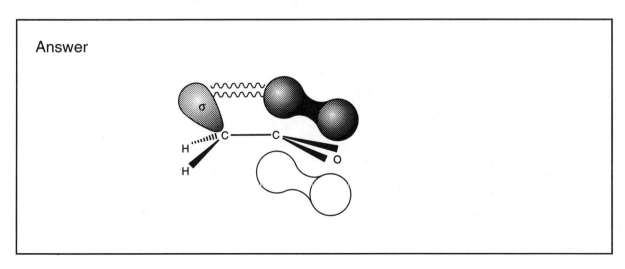

A Newman diagram can be drawn by looking along the C–C bond to indicate the relative orientation of the α hydrogen which will be lost.

Enolates formed from ketones or aldehydes are extremely important in the synthesis of more complex organic molecules. Since they are negatively charged, they are nucleophiles and can react with organic molecules which contain electrophilic centres. We shall cover two examples of this—reaction of enolates with alkyl halides and reaction of enolates with other carbonyl groups.

11.2 Reaction of enolates with alkyl halides

Enolates can react with alkyl halides.

Let us consider the reaction of the following enolate and iodomethane. Identify the electrophilic and nucleophilic centres in both molecules.

Answer

Nucleophilic centre

Weak nucleophilic centre

Electrophilic centre

The enolate has a negatively charged oxygen and this is a nucleophilic centre.

Iodomethane has an electronegative iodine which will polarise the C–I bond making the iodine electron-rich and the neighbouring carbon electron-deficient. Therefore, the iodine could act as a nucleophilic centre and the carbon could act as an electrophilic centre. From previous sections, we know that iodine is a very poor nucleophile and so iodomethane is more likely to act as an electrophile at the electrophilic carbon.

Having identified the nucleophilic centre in the enolate and the electrophilic centre in iodomethane, draw a mechanism for a likely reaction between these molecules and explain which electrons are being used in the making and breaking of bonds.

Answer

The nucleophilic centre on the enolate is the negatively charged oxygen. It will use one of its lone pairs of electrons to form a new bond to the electrophilic carbon on iodomethane. As soon as this new bond forms, one of the four bonds already round carbon must break. The weakest bond is the bond to iodine. This breaks and both electrons move onto iodine to give it a fourth lone pair of electrons and a negative charge.

(Notice that as far as the alkyl halide is concerned this is another example of nucleophilic substitution.)

This reaction is possible, but a more important product can be obtained by the formation of a C–C bond. For that to happen, the charge on the enolate ion has to be on a carbon atom.

Draw a resonance mechanism to show how the negative charge can be placed on such an atom.

Answer

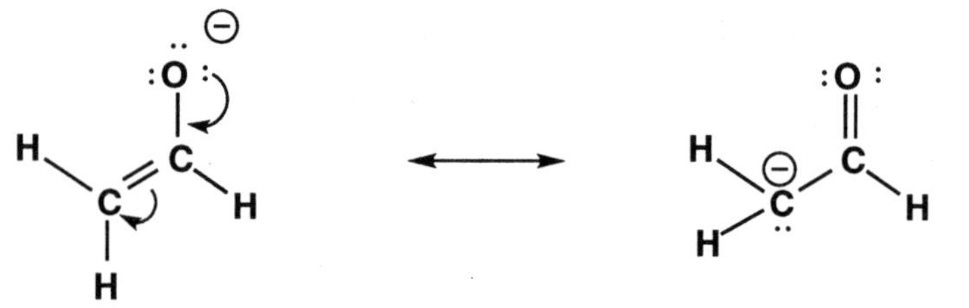

Now the negative charge is on the carbon atom, making it the nucleophilic centre. Show the mechanism by which this ion will react with iodomethane, and draw the product formed.

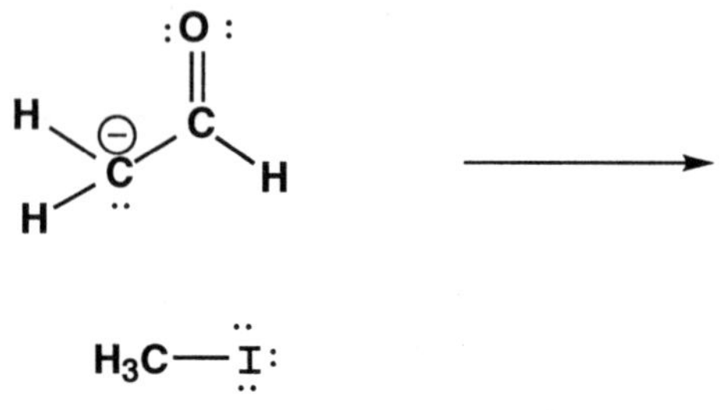

Answer

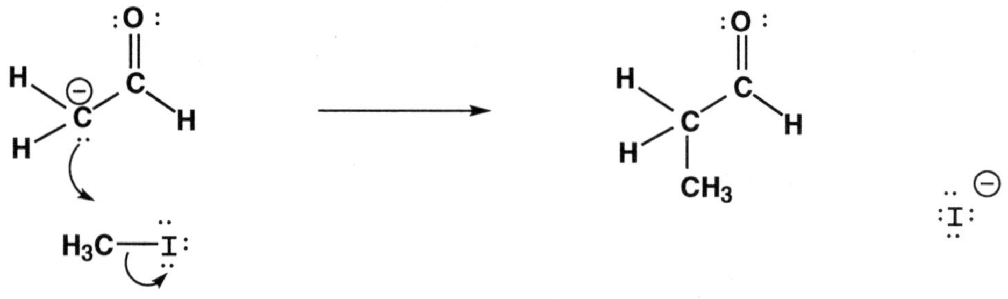

A mechanism can be drawn directly from the enolate without having to go though the resonance structure we have just drawn. See if you can work out what it is. (Hint: remember how the enolate was formed in 'one go' from the original carbonyl compound.)

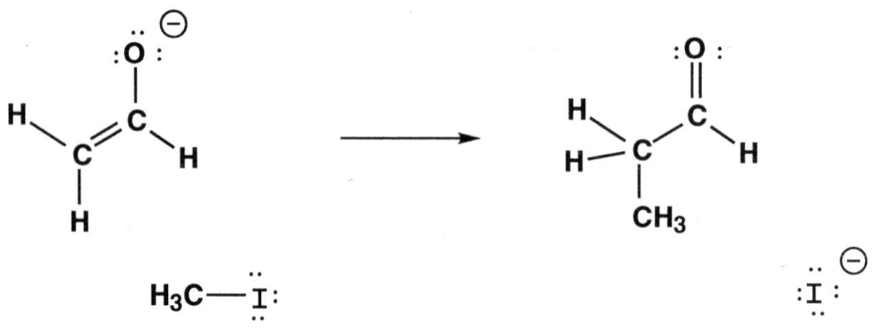

Answer

Note: as far as the alkyl halide is concerned, this is a nucleophilic substitution which will take place via the S_N2 mechanism.

The overall reaction introduces an alkyl group α to the carbonyl functional group and is a very useful reaction in organic synthesis. It is found that *C*-alkylation occurs more frequently than the alternative *O*-alkylation mentioned earlier, but this is beyond the scope of this text.

Starting from propanal, how would you synthesise the following?

Answer

The first reaction would be carried out by reacting propanal with base to give the enolate ion. This would then be treated with iodomethane to give the extra methyl group.

The second reaction could be carried out using excess base and iodomethane. The reaction would proceed exactly the same way to give the product of the first reaction. In the presence of excess base, this would be converted to another enolate ion which would then react with excess iodomethane to introduce the second methyl group.

What difficulties might you foresee in carrying out the following reaction? Are there any other products possible?

In a molecule like this, there are α protons on either side of the carbonyl group and so two different products are possible.

Answer

The reaction can be controlled to favour one or other of the possible products, but we do not have the space to cover that here.

11.3 Reaction of enolates with carbonyl compounds (the aldol reaction)

We have seen how enolates react with alkyl halides. Enolates can also react with aldehydes and ketones.

Aldehydes and ketones have an electrophilic centre. Where? What sort of reaction do they commonly undergo?

Answer

The carbon of the carbonyl group is electrophilic. Aldehydes and ketones normally undergo nucleophilic additions.

Draw a mechanism for the first stage of the reaction between the enolate below and ethanal. Remember that the enolate is the nucleophile and will react as if the carbon atom is negatively charged. Ethanal will act as the electrophile and will undergo a typical nucleophilic addition.

Answer

The reaction could now be worked up by adding water to give the final product. Show the mechanism for the final stage. (Note that the intermediate has been redrawn 'more tidily'.)

Answer

Note that the product contains two functional groups—an aldehyde and an alcohol. The reaction is called an Aldol reaction (**ald**ehyde/alcoh**ol**.)

Summary

- Hydrogens attached to carbon can be acidic if the carbon is α to a carbonyl group.

- Aldehydes and ketones bearing an α proton can react with a base to give enolate anions.

- Aldehydes and ketones bearing an α proton can be alkylated at the alpha position.

- Aldehydes and ketones bearing an α proton can react with another aldehyde or ketone by the Aldol reaction.

Elimination Reactions of Alkyl Halides and Alcohols

Elimination Reactions of Alkyl Halides and Alcohols

12.1 Elimination of alkyl halides

There is another situation in which protons can be lost from a carbon atom. Alkyl halides can lose a proton along with a halide ion to give an alkene. The lost proton is attached to a carbon 'next door' to the electrophilic carbon of the alkyl halide. In this respect, there is a similarity in the position of this susceptible proton and the position of the acidic proton relative to a carbonyl group.

In both cases, the susceptible proton is two bonds removed from an electrophilic carbon. In the case of the carbonyl group, we have seen that the electrophilic centre 'pulls-in' the electrons from the C–H bond. A C=C bond is formed and the π bond of the carbonyl groups breaks such that both π electrons move onto the oxygen atom. What do you think happens in the case of the alkyl halide? It should be possible for you to work this out for yourself, if you try to draw a similar mechanism to the one which created the enolate ion. Draw a mechanism to show a hydroxide ion 'picking-off' the susceptible proton in the following alkyl halide. Show in your mechanism the most likely destination for the electrons in the C–H bond, assuming that they do not stay on the carbon atom. Having decided that, you may have other bonds to break, so decide which is the most likely and see what is left over.

Answer

The mechanism is as follows.

As the hydroxide ion forms a bond to the susceptible proton, the C–H bond breaks. Both electrons could move onto the carbon, but there is a neighbouring electrophilic carbon which attracts the electrons and so the electrons move in to form a π bond between the two carbons. At the same time as this π bond is formed, a bond must break so that the electrophilic carbon atom does not have more than four bonds. The only likely bond is the C–I bond since all the others are strong C–C or C–H bonds. Both electrons end up on the iodine atom and this atom is lost as the iodide ion. The overall reaction is an elimination where HI is lost from the molecule and an alkene is formed.

12.2 Substitution versus elimination—primary alkyl halides

The mechanism above is a concerted process and is termed an E2 mechanism where the E stands for elimination and the 2 shows that the rate of reaction depends on the two starting materials. You may remember we discussed the S_N2 reaction earlier. If you cannot remember the reaction, go to Section 10.1 and go over it before proceeding any further on this section.

Assuming you are now familiar with the S_N2 reaction, you might be a bit puzzled! According to Section 10, the hydroxide ion attacks the electrophilic carbon of an alkyl halide and pushes off the halogen to give an alcohol by a nucleophilic substitution reaction. However, we have just been talking about alkyl halides reacting with the hydroxide ion by a totally different mechanism (elimination) to give a totally different product (an alkene). Which of these reactions is correct?

The answer is that both are correct. Many organic molecules can undergo different reactions with the same reagent. Thus, organic molecules containing an alkyl halide functional group can react by a substitution reaction and an elimination reaction to give a mixture of products. Analyse the reactants in the following reaction and explain what products you might expect.

$$H_3C$$
$$\text{CH—CH}_2\text{-I} \quad + \quad \overset{\oplus}{Na} \quad \overset{\ominus}{OCH_2CH_3} \quad \longrightarrow$$
$$H_3C$$

Answer

NaOEt is a salt made up of a sodium cation and an ethoxide anion. The ethoxide ion is a nucleophile and will therefore form a bond to an electrophilic centre on the alkyl iodide. The most obvious electrophilic centre on the alkyl halide is the carbon atom next to the iodine atom. The iodine atom is electronegative and has a greater share of the electrons in the C–I bond, making the carbon slightly positive. The reaction would be a nucleophilic substitution and give an ether.

$$H_3C \quad \overset{\delta+}{} \quad \overset{\delta-}{} \quad \overset{NaOEt}{\longrightarrow} \quad H_3C$$
$$\text{CH—CH}_2\text{—I} \qquad\qquad \text{CH—CH}_2\text{—OCH}_2CH_3 \quad + \text{ NaI}$$
$$H_3C \qquad\qquad\qquad\qquad\qquad\qquad H_3C$$

However, the alkyl halide can also undergo an elimination reaction if the ethoxide ion acts as a base and forms a bond to the susceptible proton. The product is an alkene.

$$H_3C \quad \overset{\delta+}{} \quad \overset{\delta-}{}$$
$$H\text{——}C\text{—CH}_2\text{-I} \quad \overset{NaOEt}{\longrightarrow} \quad H_3C \qquad H$$
$$\qquad H_3C \qquad\qquad\qquad\qquad\qquad C{=}C \qquad + \text{ NaI } + \text{ EtOH}$$
$$\qquad\qquad\qquad\qquad\qquad\qquad H_3C \qquad H$$

susceptible
proton

Generally, primary alkyl halides prefer to react with nucleophiles by the S_N2 mechanism to give substitution products. Elimination can also occur by the E2 mechanism to give smaller quantities of alkene. The ratio of substitution and elimination products depends on the type of nucleophile used and this is beyond the scope of this text.

12.3 Substitution versus elimination—tertiary alkyl halides

We have looked at primary alkyl halides and seen that the elimination reaction is usually less important than substitution. Let us look now at tertiary alkyl halides. What do you recall about the reaction of tertiary alkyl halides with nucleophiles from Section 10?

Answer

Tertiary alkyl halides react with nucleophiles by the S_N1 reaction via a carbocation intermediate. The electrophilic centre is shielded by alkyl substituents from the incoming nucleophile and the halide ion has to depart first to give a planar carbocation before there is room for the incoming nucleophile.

Show the S_N1 mechanism for the following reaction and suggest how an elimination reaction could occur via the same carbocation intermediate to give an alkene.

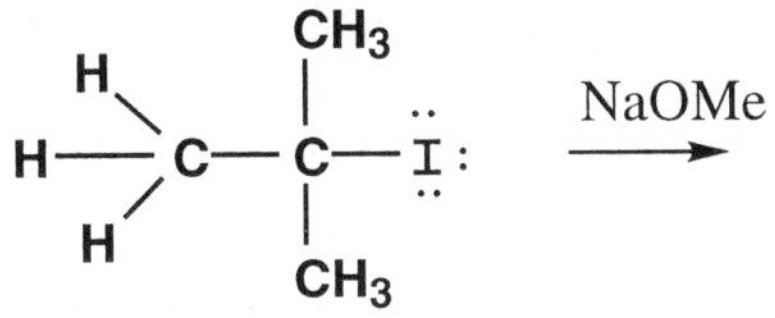

Nucleophilic substitution (S_N1)

Elimination

Answer

The alkyl iodide has a polar C–I bond making the carbon atom electrophilic. The methoxide ion is nucleophilic but is unable to attack the electrophilic carbon of the alkyl halide due to the shielding effect of the alkyl groups surrounding the electrophilic carbon. The C–I bond can break to ease the steric strain in the molecule and give a carbocation where the positive charge is stabilised by the electron-donating effect of the three alkyl groups surrounding it, as well as by hyperconjugation. The methoxide ion acts as a nucleophile and forms a bond to the carbocation to give an ether product.

cont.

The carbocation can also react with the methoxide ion by the alternative elimination mechanism below to give the alkene product.

This mechanism is termed an E1 mechanism. The 1 indicates that the rate of reaction is dependent only on the concentration of alkyl halide present in the reaction. Once again, two different products are possible from this reaction, but unlike primary alkyl halides, the preferred product tends to be the elimination product rather than the substitution product. Suggest why this might be so.

Answer

The positively charged carbon atom is the obvious place for an attacking nucleophile to attack. However, the carbocation was formed in the first place in order to relieve the strain of the crowded alkyl iodide. If the nucleophile forms a bond to the carbocation centre, the tetrahedral centre is reformed and the three alkyl groups are crowded together again. On the other hand, if the nucleophile goes for the susceptible proton, the product is the planar alkene and there is no crowding. Therefore, a more stable product is formed. It is also worth noting that the susceptible proton is more accessible to the nucleophile compared to the carbocation centre which is in the heart of the molecule.

By varying reaction conditions, it is possible to alter the ratio of elimination and substitution products but that is beyond the scope of this text.

12.4 Substitution versus elimination—secondary alkyl halides

With secondary alkyl halides, it is difficult to predict whether the substitution or the elimination product will be preferred, and it is likely that a mixture of both will be formed. The type of mechanism will also depend on the structure of the alkyl halide and the nature of the nucleophile. This is beyond the scope of this text.

12.5 Elimination of alcohols

The oxygen of an alcohol functional group is electronegative and so by analogy with alkyl halides, elimination reactions should be possible.

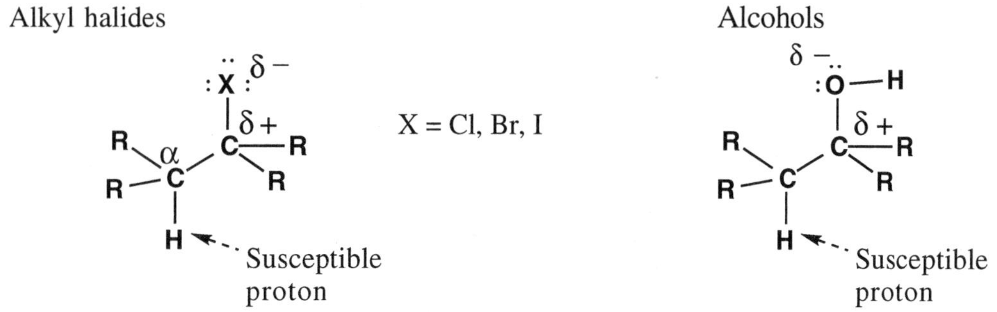

Alkyl halides   Alcohols

X = Cl, Br, I

In fact, the reaction does not take place very easily. Can you suggest why?

Answer

Oxygen is less electronegative than a halogen. Therefore, the carbon atom bound to the alcohol group is not so electrophilic. Secondly the required leaving group (OH) is a poorer leaving group than a halide (X^-) and so elimination is more difficult.

It is possible to form alkenes from alcohols under acid conditions. The acid conditions are important and it is found that tertiary alcohols react more easily than secondary alcohols, which in turn react more easily than primary alcohols.

The following reaction is an example of the elimination reaction of tertiary alcohols.

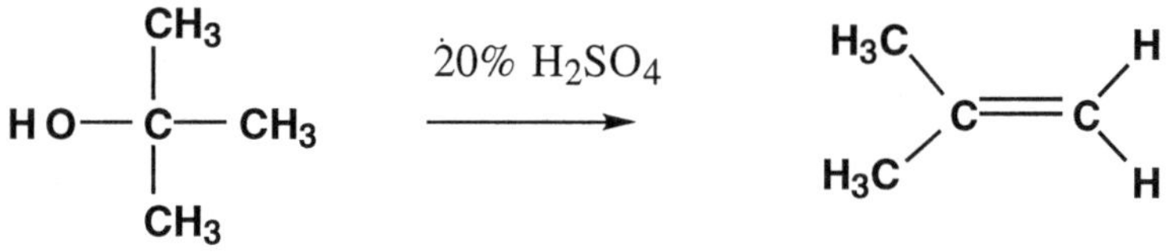

What molecule has been lost from the alcohol?

Answer

Water has been lost from the molecule.

Since water has been lost from the molecule, this reaction is also known as a **dehydration** reaction. Circle the atoms which were lost from the alcohol to form water.

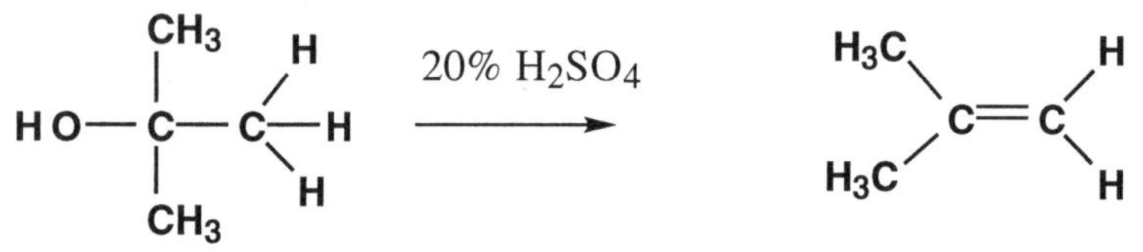

Answer

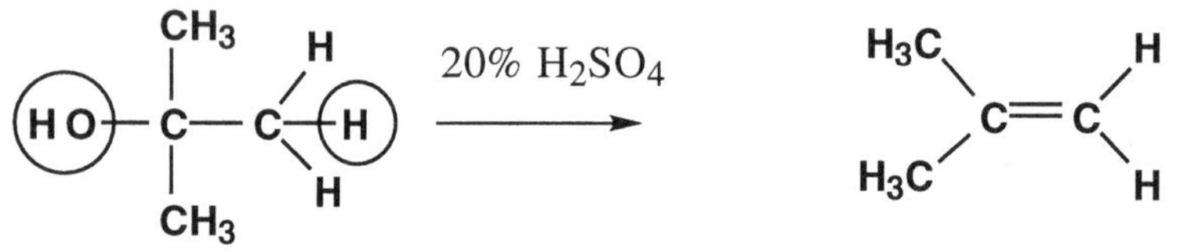

The HO group is lost along with a hydrogen atom from one of the CH_3 groups.

We are now ready to look at the mechanism. We are expecting an E1 mechanism but we need a nucleophile (or base) for this mechanism. The reaction conditions are acidic and so where are we going to get the necessary nucleophile? The reaction conditions are 20% H_2SO_4. What does the 20% mean?

Answer

The 20% means that conc. H_2SO_4 has been diluted with water such that the solution is 20% H_2SO_4 and 80% water.

Is there any molecule which could act as a nucleophile under these reaction conditions?

Answer

Water itself can act as the nucleophile.

We shall now look at the mechanism in detail. We have acidic conditions, so the most likely reaction to start with is the reaction of the alcohol with a proton. Identify the nucleophilic and electrophilic centres in both species and draw a mechanism to show the most likely reaction.

Answer

The proton is positively charged and is an electrophile. It will therefore react with a nucleophilic centre on the alcohol. The only nucleophilic centre on the alcohol is the oxygen atom. The oxygen atom uses one of its lone pairs of electrons to form an O–H bond. By doing so, it effectively loses an electron since it now has to share two electrons in the new bond. Therefore the oxygen gains a positive charge.

It is now possible to form the carbocation necessary for the E1 mechanism. Draw this mechanism and suggest why protonation prior to this step is advantageous.

Answer

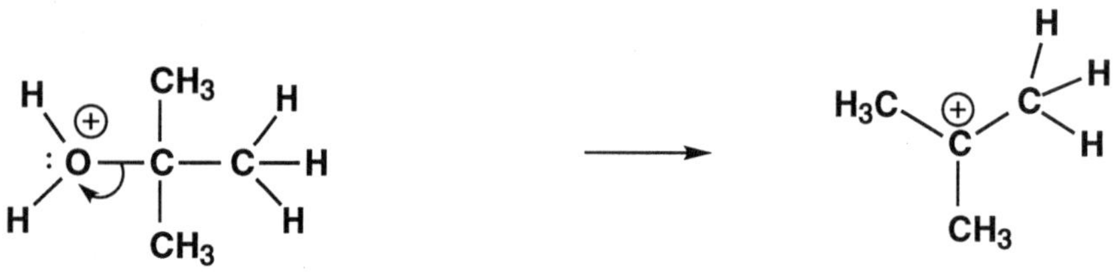

Protonation generates a positively charged oxygen which makes the C–O bond easier to break since the electrons moving onto oxygen neutralise that charge. Water is a far better leaving group than the hydroxide ion.

Water is the nucleophile present in solution. Show a mechanism which gives the final alkene.

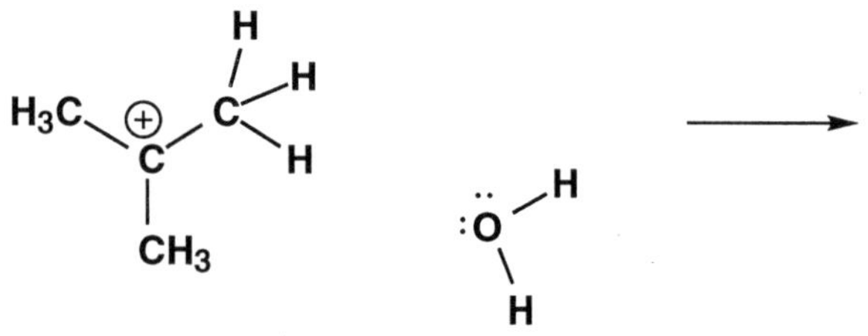

Answer

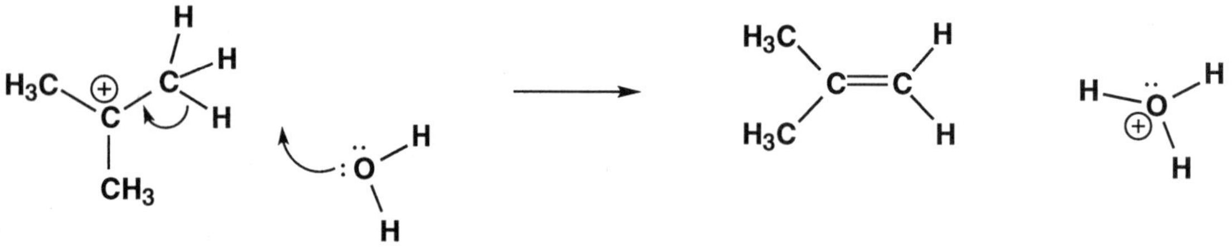

The nucleophilic centre of water is the oxygen atom. A lone pair of electrons is used to form a bond to the susceptible proton of the carbocation. At the same time, the C–H bond is broken and both electrons are used to form a π bond between two carbons.

Summary

- Alkyl halides bearing a susceptible proton can react with a nucleophile by nucleophilic substitution and by elimination.

- Elimination proceeds fastest for tertiary halides, followed by secondary halides then primary halides.

- Primary alkyl halides usually prefer to react with nucleophiles by nucleophilic substitution (S_N2 mechanism). Elimination may also occur by the E2 mechanism.

- Tertiary alkyl halides usually prefer to react with nucleophiles by elimination (E1 mechanism) to give alkenes. Some nucleophilic substitution may take place via the S_N1 mechanism.

- It is not possible to predict the reaction of secondary alkyl halides with any confidence. A mixture of substitution and elimination products is likely.

- Alcohols can undergo the elimination reaction (dehydration) under acidic conditions.

SECTION 13

Reductions and Oxidations

Reductions and Oxidations

13.1 Oxidation levels

Reductions and oxidations are commonly used to convert one functional group into another. It is therefore useful to know which functional groups can be reduced and which functional groups can be oxidised.

The first thing to appreciate is that different functional groups have different oxidation levels and by knowing these, you will know whether the conversion between two functional groups involves an oxidation or a reduction.

We can group the functional groups into the following four oxidation levels, where level 1 is the least oxidised (or the most reduced) level.

OXIDATION

Level 1	Level 2	Level 3	Level 4
Alkanes	Alkenes Alkyl halides Alcohols	Alkynes Aldehydes Ketones	 Carboxylic acids Acid derivatives
	Amines Ethers	Imines	Nitriles

REDUCTION

From the above table, determine whether the following reactions are oxidations, reductions or neither.

(a)  $CH_3 - C \equiv C - CH_3$ ⟶ (2-butene with H, H on one side and CH₃, CH₃) ▭

(b) $CH_3CH_2 - OH$ ⟶ (acetaldehyde H_3C, H, $C=O$) ▭

(c) (acetaldehyde H_3C, H, $C=O$) ⟶ (acetic acid H_3C, OH, $C=O$) ▭

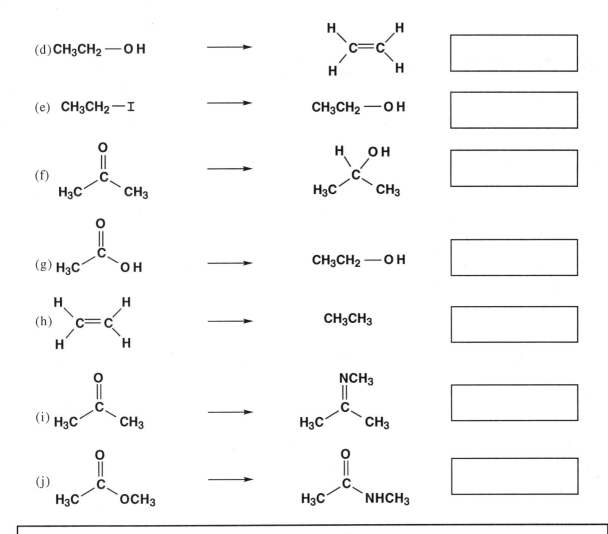

Answer

Reactions (a), (f), (g) and (h) are reductions where a functional group in a higher oxidation level is being converted to a functional group in a lower oxidation level.

Reactions (b) and (c) are oxidations since a functional group in a lower oxidation level is being converted to a higher oxidation level.

Reactions (d), (e), (i) and (j) are neither oxidations nor reductions since the functional groups involved are at the same oxidation level.

13.2 Reducing agents

There are many reducing agents available to organic chemists. The following are some of the more important reducing agents and the reductions which can be carried out with them.

- $LiAlH_4$

    Reduces ketones to secondary alcohols, aldehydes to primary alcohols.

    Reduces carboxylic acids, acid chlorides, acid anhydrides and esters to primary alcohols.

    Reduces amides to amines.

- $NaBH_4$

    Reduces ketones to secondary alcohols and aldehydes to primary alcohols.

- $LiAlH(OBu^t)_3$

    Reduces acid chlorides to aldehydes.

- H_2 with a palladium or nickel catalyst

   Reduces alkenes and alkynes to alkanes.

   Reduces nitriles to amines.

- H_2 with a poisoned catalyst (Lindlar's catalyst)

   Reduces alkynes to (Z)-alkenes.

- Na or Li/NH_3

   Reduces alkynes to (E)-alkenes.

- Zn/HCl or NH_2NH_2/NaOH

   Reduces aldehydes and ketones to alkanes.

- Sn/HCl

   Reduces nitro groups to amines.

Which reducing agent would you use for the following reductions?

(a) 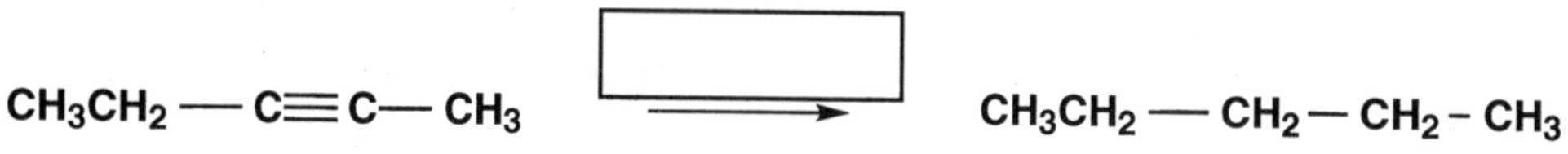 CH_3CH_2—C≡C—CH_3 ⟶ CH_3CH_2—CH_2—CH_2–CH_3

(b) 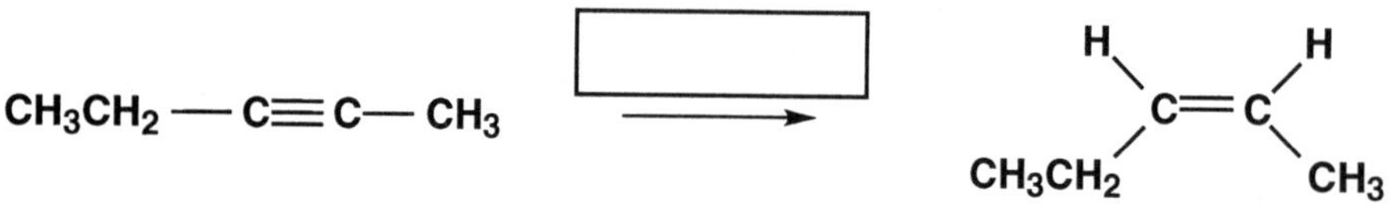 CH_3CH_2—C≡C—CH_3 ⟶

$$\underset{CH_3CH_2}{\overset{H}{\diagdown}}C=C\underset{CH_3}{\overset{H}{\diagup}}$$

(c) CH_3CH_2—C≡C—CH_3 ⟶

$$\underset{H}{\overset{CH_3CH_2}{\diagdown}}C=C\underset{CH_3}{\overset{H}{\diagup}}$$

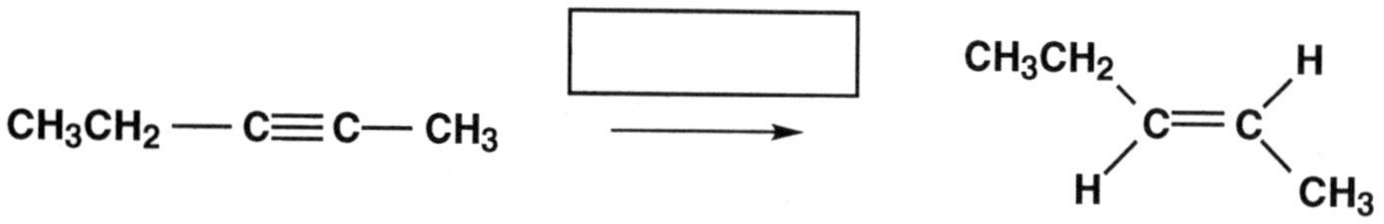

(d)

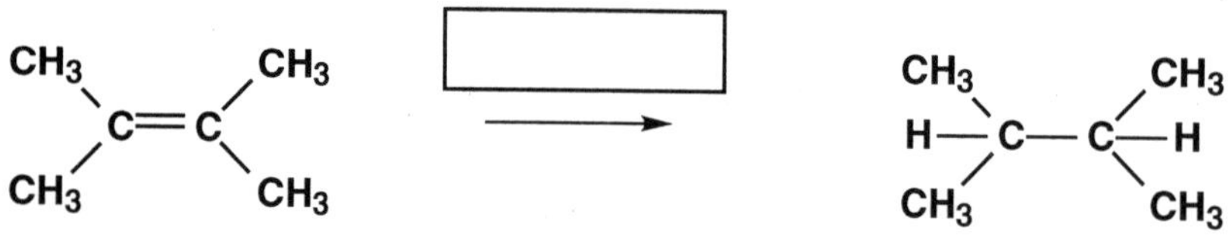

$$\underset{CH_3}{\overset{CH_3}{\diagdown}}C=C\underset{CH_3}{\overset{CH_3}{\diagup}} \longrightarrow \underset{CH_3}{\overset{CH_3}{\mid}}H—C—C—H\underset{CH_3}{\overset{CH_3}{\mid}}$$

(e) 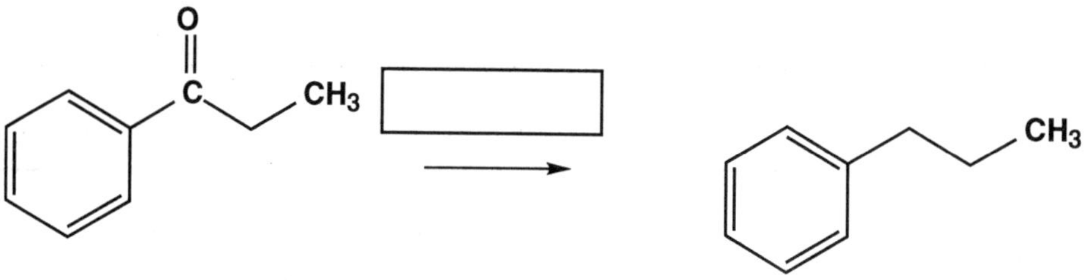 ⟶

(f) 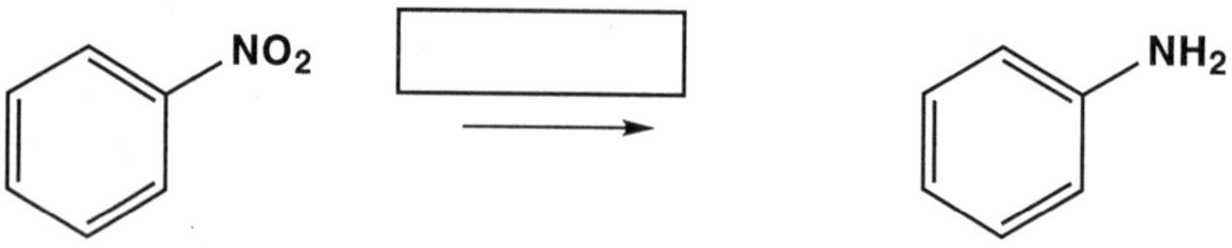 NO_2 ⟶ NH_2

(g)

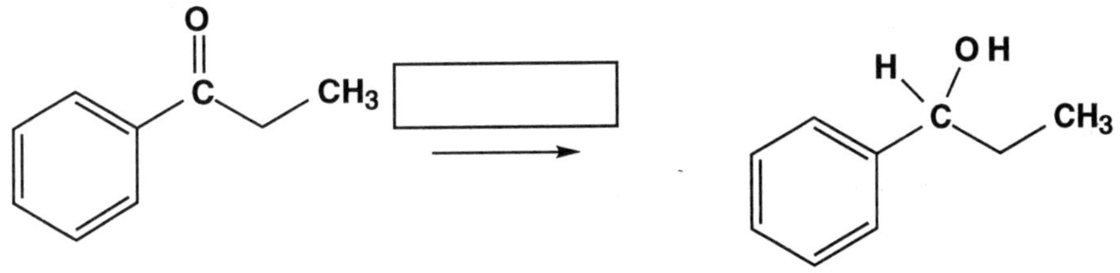

(h)

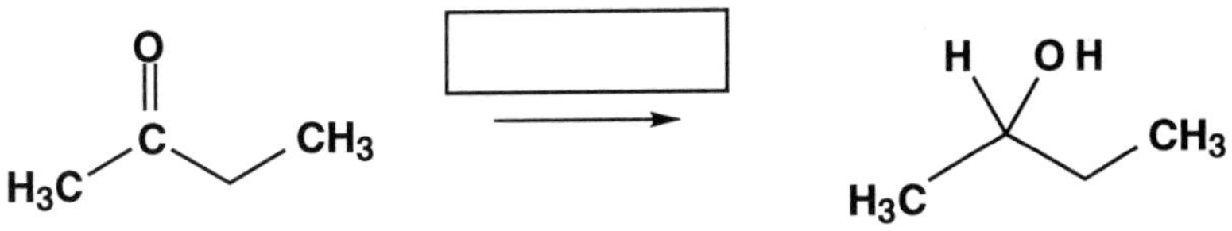

(i)

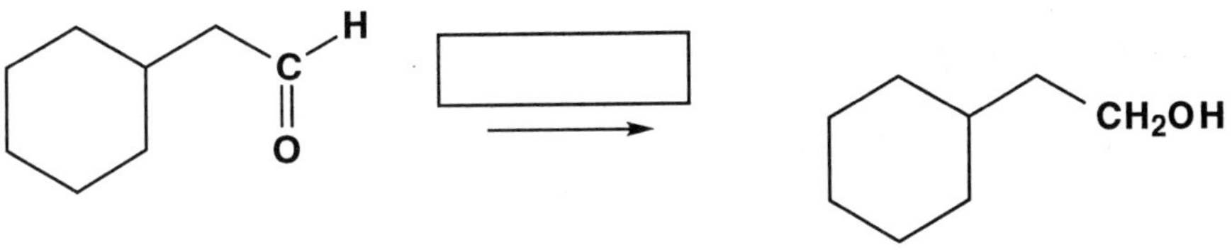

(j)

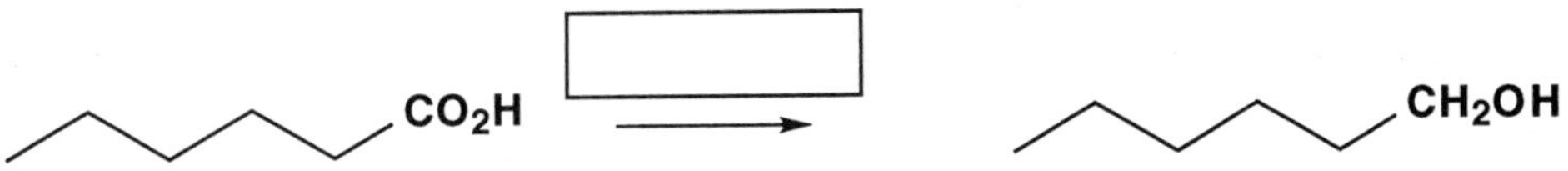

(k)

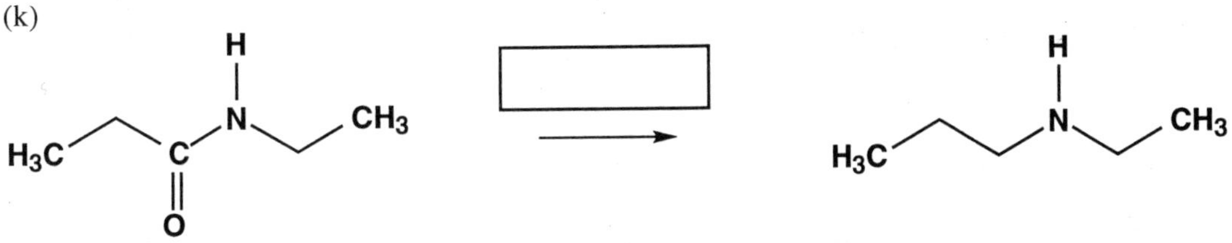

Answer

The reducing agents for each reaction are:

(a) H_2 with Pd or Ni catalyst   (b) H_2 with Lindlar's catalyst   (c) Na or K/NH_3

(d) H_2 with Pd or Ni catalyst   (e) Zn/HCl or N_2H_4/NaOH   (f) Sn/HCl

(g) NaBH$_4$ or LiAlH$_4$ (h) NaBH$_4$ or LiAlH$_4$ (i) NaBH$_4$ or LiAlH$_4$

(j) LiAlH$_4$ (k) LiAlH$_4$

Knowing the reducing properties of a reducing agent is important if you wish to reduce one functional group in a molecule without reducing another.

Which reducing agents would you choose to carry out the following reactions?

(a)

(b)

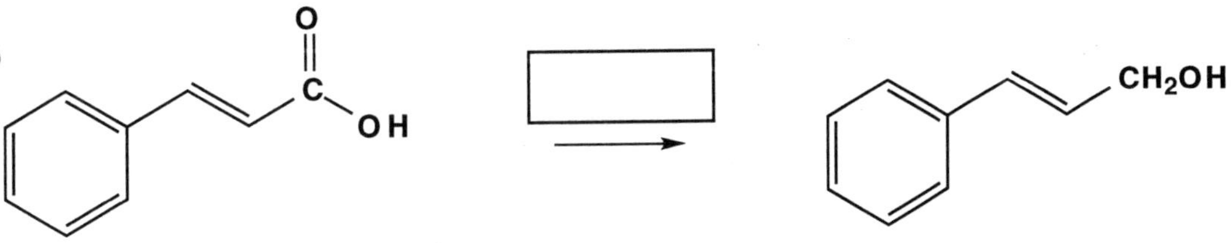

Answer

For reaction (a) the alkene is to be reduced and the carboxylic acid is to be left alone. H_2 over a palladium/charcoal catalyst would achieve this.

For reaction (b) $LiAlH_4$ would be the reducing agent of choice, since it would reduce the acid and leave the alkene alone.

$NaBH_4$ would be useless for either reaction since it is specific for ketones and aldehydes.

Identify the reducing agents which should be used for the following reactions.

(a)

(b)

Answer

$NaBH_4$ would be the reducing agent for reaction (a) and $LiAlH_4$ would be the reducing agent for reaction (b).

13.3 Oxidising agents

The following oxidising agents are commonly used.

- Potassium dichromate ($K_2Cr_2O_7$)

    Oxidises primary alcohols to carboxylic acids, secondary alcohols to ketones, aldehydes to carboxylic acids and alkylbenzenes to aromatic carboxylic acids.

- Potassium permanganate ($KMnO_4$)

    Oxidises primary alcohols to carboxylic acids, alkenes to 1,2-diols, alkylbenzenes to aromatic carboxylic acids and aldehydes to carboxylic acids.

- Pyridinium chlorochromate

    Oxidises primary alcohols to aldehydes.

- Peracids (R—C—O—OH)

    Oxidises alkenes to epoxides.

- Osmium tetraoxide (OsO_4)

    Oxidises alkenes to 1,2-diols.

- Osmium tetraoxide (OsO_4) followed by periodic acid (HIO_4)

    Oxidises alkenes to aldehydes and ketones.

- Ozone plus Zn/H_2O or Me_2S work up.

  Oxidises alkenes to aldehydes and ketones.

Which reagents would you use for the following oxidations?

(a)

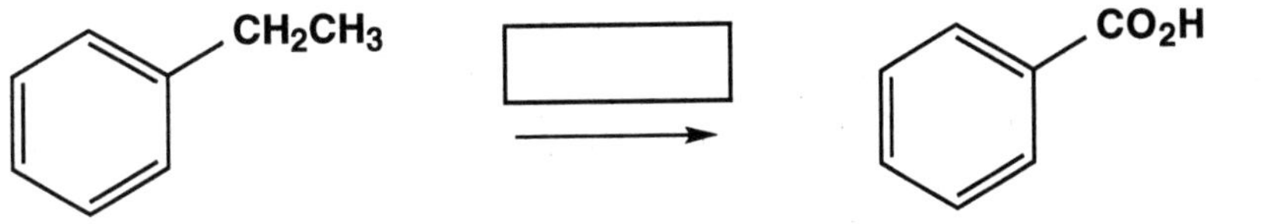

(b)

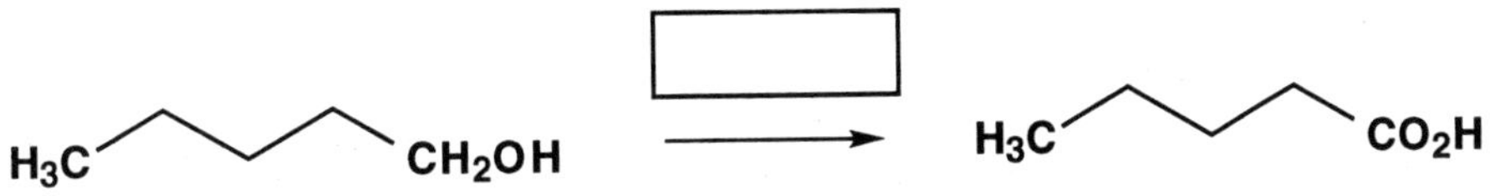

(c)

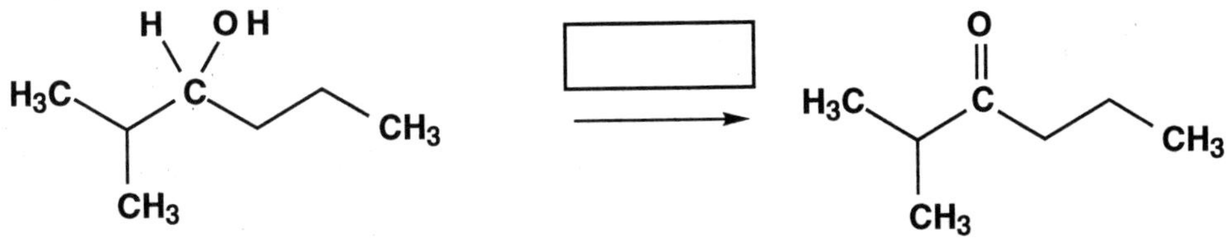

(d)

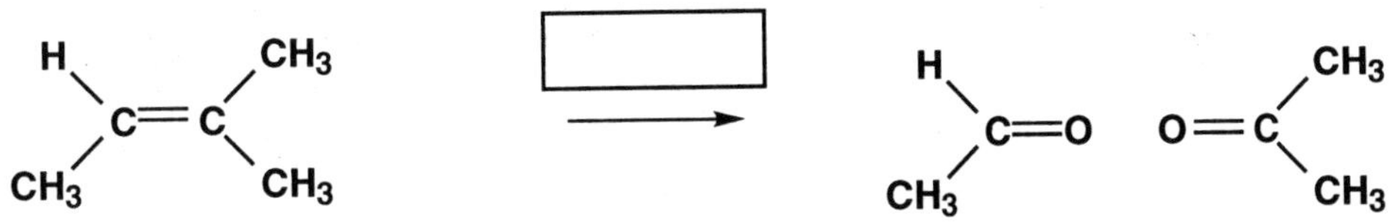

(e)

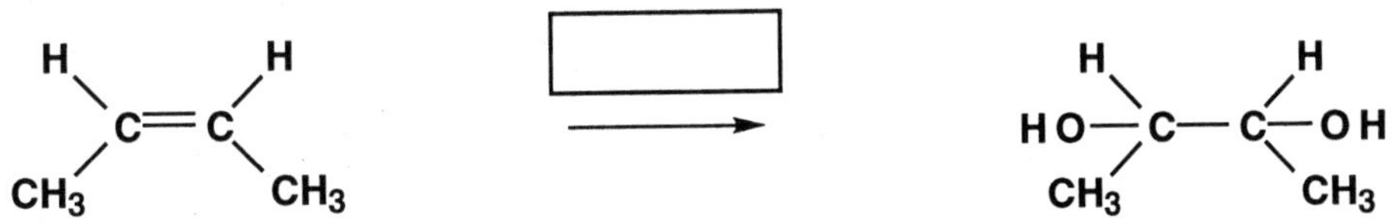

(f)

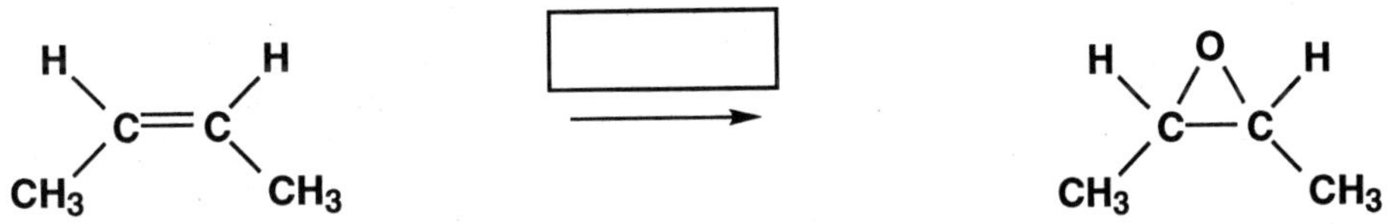

Answer

The oxidising agents which could be used are as follows:

(a) $KMnO_4$ or $K_2Cr_2O_7$ (b) $KMnO_4$ or $K_2Cr_2O_7$ (c) $KMnO_4$ or $K_2Cr_2O_7$

(d) O_3 and Zn/H_2O (e) $KMnO_4$ or OsO_4 (f) RCO_2OH

   or OsO_4 then HIO_4

SECTION 14

Radical Reactions

Radical Reactions

14.1 Free radical halogenation of alkanes

The reactions and mechanisms which we have studied so far involve the movement of pairs of electrons to form new bonds or new lone pairs of electrons. These are known as heterolytic mechanisms—where both electrons are moving together.

The free radical halogenation of alkanes involves a different type of reaction mechanism—where the electrons do not move in pairs. This sort of reaction is called a homolytic reaction where the two electrons in a bond move in different directions.

The reaction we are going to look at is the halogenation of an alkane. Alkanes are very unreactive molecules but if you treat an alkane with chlorine in the presence of light or heat, a reaction can take place and an alkyl halide is formed.

$$H_3C\!-\!H \quad + \quad Cl\!-\!Cl \quad \xrightarrow{\text{Light or heat}} \quad H_3C\!-\!Cl \quad + \quad H\!-\!Cl$$

The light or the heat is essential for this reaction to work. Light or heat provides energy to split the chlorine molecule into individual atoms. Unlike the reactions we have looked at so far, the electrons in the Cl–Cl bond do not move as a pair. Instead, they each move onto a separate chlorine. In other words, one electron moves one way, while the other moves the opposite way.

$$:\!\overset{..}{\underset{..}{Cl}}\!-\!\overset{..}{\underset{..}{Cl}}\!: \quad \xrightarrow{\text{Light or heat}} \quad :\!\overset{..}{\underset{..}{Cl}}\!\cdot \quad \cdot\overset{..}{\underset{..}{Cl}}\!:$$

We can show the mechanism for this by using half arrows. The half arrow indicates the movement of one electron. Use half arrows to show the mechanism for the splitting of the Cl–Cl bond. Remember that the electrons are in the bond and so the half arrows must start from the centre of the bond.

$$:\!\overset{..}{\underset{..}{Cl}}\!-\!\overset{..}{\underset{..}{Cl}}\!: \quad \xrightarrow{\text{Light or heat}} \quad :\!\overset{..}{\underset{..}{Cl}}\!\cdot \quad \cdot\overset{..}{\underset{..}{Cl}}\!:$$

Answer

$$:\overset{..}{\underset{..}{Cl}} \frown \overset{..}{\underset{..}{Cl}}:\quad \xrightarrow{\substack{\text{Light or} \\ \text{heat}}} \quad :\overset{..}{\underset{..}{Cl}}\cdot \qquad \cdot\overset{..}{\underset{..}{Cl}}:$$

The chlorine atoms which are obtained are neutral since they have seven valence electrons. However, since they have an unpaired valence electron, they are very reactive and are known as **radicals**. This first step is known as a chain-initiating step since the generation of chlorine radicals initiates a chain reaction. Chlorine radicals are reactive enough to remove a hydrogen atom from an alkane and give a methyl radical.

$$H_3C{-}H \quad + \quad \cdot\overset{..}{\underset{..}{Cl}}: \quad \longrightarrow \quad H_3C\cdot \quad + \quad H{-}\overset{..}{\underset{..}{Cl}}:$$

We need to identify what has happened to the electrons in this process. A C–H bond is cleaved. Where do the electrons end up?

Answer

There are two electrons in the C–H bond. The methyl radical takes one of these electrons. The other one is used along with the single electron provided by the chlorine radical in the formation of the H–Cl bond.

Show the mechanism for this. Once again you will have to use half arrows to show the movement of single electrons.

$$H_3C{-}H \quad \cdot\overset{..}{\underset{..}{Cl}}: \quad \longrightarrow \quad H_3C\cdot \quad H{-}\overset{..}{\underset{..}{Cl}}:$$

Answer

$$H_3C \frown H \quad \frown \quad \cdot\overset{..}{\underset{..}{Cl}}: \quad \longrightarrow \quad H_3C\cdot \quad H{-}\overset{..}{\underset{..}{Cl}}:$$

Notice that another radical has been created. This step is known as a chain propagation step since one radical has been used up and another one has been created.

Another two propagation steps are possible where the methyl radical reacts with a molecule of methane or with a molecule of chlorine. Show the mechanisms by which these steps take place.

H₃C ·   : C̈l — C̈l :   ⟶   H₃C — C̈l :   · C̈l :

H₃C ·   H — CH₃   ⟶   H₃C — H   · CH₃

Answer

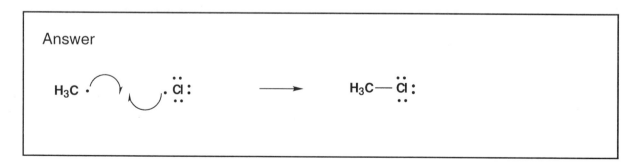

There are still chlorine radicals around and so it is possible for a methyl radical and a chlorine radical to meet each other and form a bond. Show a mechanism by which this can take place.

H₃C ·   · C̈l :   ⟶   H₃C — C̈l :

Answer

H₃C ·   · C̈l :   ⟶   H₃C — C̈l :

This is known as a chain termination step. Why?

Answer

This reaction removes two radicals. No more radicals are produced to continue the chain reaction.

Another chain termination reaction is possible involving two methyl radicals. Draw the mechanism for this and give the products obtained.

H₃C ·   · CH₃   ⟶

Answer

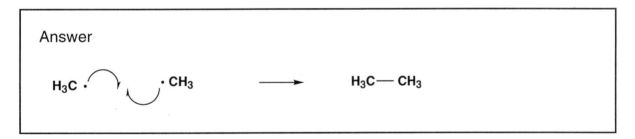

The chain reaction will continue until all the radicals have been used up. Since the radicals are so reactive, they are not very specific and a mixture of structures can be obtained. How many different alkyl halides can be obtained from the chlorination of hexane and of cyclohexane? Which reaction is likely to be the most useful?

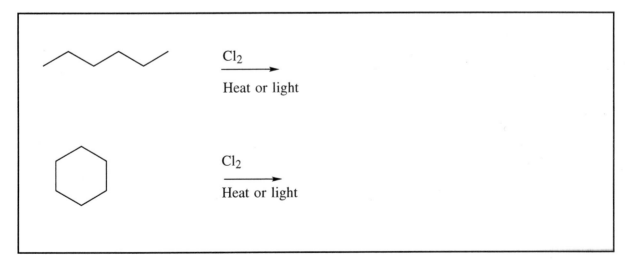

Answer

Assuming that only monochlorination takes place, the chlorination of hexane could give three different products.

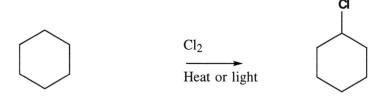

The chlorination of cyclohexane will give only one product and is therefore a more useful reaction.

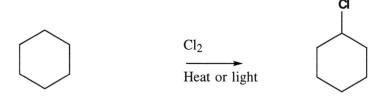